Up Yours

What to do With an Engineering, Research, and Innovation Career

Clifford L Spiro, PhD

ISBN: 978-1-4834-0303-8 (sc)
ISBN: 978-1-4834-0302-1 (e)

Lulu Publishing Services rev. date: 08/21/2013

This book is dedicated to the hundreds of colleagues including peers, employees, managers, customers, and suppliers that I have had the privilege to serve over the years. Thank you for everything.

Table of Contents

Introduction

FOR A LOT OF PEOPLE, A job is just that—a way to make a living. It is sure nice if the job is fun, easy, and pays well. People who think like this mainly work to live. They may have a sequence of jobs, some of which are interrelated, but continuity is lacking and the trajectory is choppy.

Don't get me wrong—my wise old landscaping boss Henry Stone told me forty years ago, "There is no honest work to be ashamed of," and I strongly agree. With all the crooks out there—tax cheats and sleazy politicians, ambulance chasing attorneys, athletes on steroids, cops on the take . . . give me an honest gardener any day. Henry, a truly thoughtful and humble man, had four sons, each one of whom became valedictorian at our local high school and undoubtedly went on to stellar careers. And though Henry lacked what I would define as a formal career, who am I to suggest that he was any better or worse off, happier or more satisfied than you, me, or the next guy?

But since you here with me right now and are probably an innovation professional or will be soon, I suspect that you are somewhat more inclined toward a career, some continuity, and a trajectory that you are hoping to accelerate by reading this book. No matter what career stage you are currently in. I truly believe that the few hours that we share in this text will be rewarding for you. At least that is my goal and hope.

Now you may now be wondering what qualifies me to be giving career counsel to techies. This is a fair question that I will briefly address now. I also hope that you will leaf ahead in the text, see a few pearls that you can put to use right away, and decide to buy the book, read it cover to cover, recommend it to all your friends and family, and refer to it often. I know—dream on, Spiro!

So, by way of a brief bio, at 59 I am a seasoned veteran of the career world. As a kid and throughout college, I worked dozens of jobs—flipping burgers, washing dishes, waiting tables, landscaping, making rubber in a unionized factory (United Cork, Rubber, Plastic, and Linoleum Workers, AFL-CIO local 3), digging swimming pools by hand . . . the usual gamut of odd jobs that kids of my time experienced. I learned a lot about people and organizations and work from these jobs, though I would say my formal career began when I completed my PhD at Caltech in 1980 and started my first real job at GE's central research laboratories. I spent the next thirty-three years working in research, development, engineering, and leading the business of technology advancement. The last ten years I spent as a VP and corporate officer. I also served on four corporate boards of directors, in two fire departments and on a board of fire commissioners, and have served on several academic and community boards. I have been writing ("R&D is War—and I've Got the Scars to Prove it"), public speaking, consulting, career-coaching, and even expert-witnessing since leaving the corporate world. In other words, I have been through the school of hard knocks from a career standpoint. I observed a lot, and have often written and spoken about careers and people. I will draw lots of examples from these experiences. Let me assure you that you will not need to be an accomplished scientist or engineer or tech-savvy in any way to read and enjoy and grow from this book. In many ways, my career experiences are the same as yours are, or will be.

I will say that things today are a lot different from when I was going through it. Some of the biggest changes I experienced included globalization, feminization, and all the benefits and challenges that the information age brought along.

Lest I risk dating myself, my first year in college was the year that the electronic calculator was introduced by Hewlett-Packard. It was so expensive that it was considered an unfair advantage to the wealthier kids, so calculators were banned and slide rules were mandated. By the time I was a sophomore just a year later, calculators were already required. My first computer science class required paper punch cards that had to be manually typed and sorted, and Fortran was the language of choice for us techies. Programs were queued and run in batch, often taking hours to get output from an overloaded IBM 360 mainframe computer. The personal computer was a mere gleam in the eye of a youngster named Steven Jobs. By the time I started graduate school, we had mini-computers the size of a refrigerator. Four years later when I started my first job, we had remote computer terminals, and a few of us could actually communicate on something called ARPA-net—the earliest version of what would become the world-wide-web. A couple years later, the miracle personal computer showed up with 8k of RAM and a whopping ten megabytes of memory. Of course you know how fast information technology (IT) and the web have progressed since then. Whole classes of jobs and careers have been created, altered, outsourced, or obliterated by IT in the name of an unending drive for productivity. If it can be automated or supplanted by an expert system, it either has been done, or will be.

Also, when I started my career, women were certainly present in science and business, but not in a large proportion. It was still rather normal that the husband worked and the wife stayed home and raised a family. Women as managers, officers, and directors were quite rare. Now, I daresay that I believe there is more nearly gender-parity though we still certainly and sadly have a ways to go. Yes women have to work harder and smarter to get noticed and move up, though less so now than just a decade or two ago. And there still is a glass ceiling at the very heights of most organizations, though I suspect that if you are cracking up against the glass, you are probably beyond reading this book, and should be writing your own.

And finally, the effects of globalization on career dynamics have been overwhelming. In *The World is Flat*, Thomas Friedman brilliantly articulated how globalization and information technology have intertwined to change virtually everything—including yours and my career. He reminded us that in my generation, it was better to be a B-student in Poughkeepsie than a superstar from Shanghai or Mumbai. Now, you better be a star no matter where you live, and that is just the ante to get into the game. It takes a lot more than talent to succeed these days, as perhaps half of the recent college graduates still seeking their first career job will attest from the bedrooms they grew up in their parents' homes.

And speaking of Shanghai and Mumbai, not only are most major corporations establishing and driving research and technology labs all over the world, especially in India and China, but even within America and Europe, more and more technical employees are of Asian origin. There are plenty of good reasons for this, mostly driven by the large number of Asian students that populate the leading graduate schools in the world. But as good students, researchers, and developers as the Asians are, I found during a seminar *"Career Decisions in American Industrial R&D Organizations"* that I gave at the Great Lakes Chinese American Chemical Society annual meeting, much of the audience was clueless regarding the career growth/career-changing pathways that were open to them. I know that those of you that were brought up in different cultures will find this book especially invaluable.

Oh, and just to make matters even tougher, in case you hadn't noticed, the world's economies are in a severe upheaval—something that I expect to be the new normal and not something anybody can just wait out.

What this all adds up to is that technical careers these days are tough, tough, tough—a lot tougher than what I went through to be sure. But that doesn't mean you can't or won't come out a winner—i.e. someone who is doing fulfilling work that gives you joy, satisfaction, and growth;

someone who has expanding opportunities including advancement and promotion with a great future; and someone who is well-compensated and can afford to live well for now and in the future.

I won't promise you that this will be easy and effortless. If you think that this book will be "Ten Simple Steps to Becoming CEO in the Next Five Years," I'm afraid you will be disappointed. But nothing we will discuss in the following chapters is unreasonable or beyond the realm of someone who has modest intelligence and is willing to work at it. And frankly, if you are already an engineer or scientist, I know you possess the brain power and dedication necessary to succeed. You probably just need to know the inner system and how to work it to grow and advance.

A career is really a big part of who you are, and it is important to make the most out of it. So let's get started on yours.

Chapter One

The Truths and Myths About Career Planning

DO YOU THINK YOU CAN PLAN your career? You can try. Sometimes, I feel that life is so random and stochastic and dynamic that maybe it is the careers that plan people, not the other way around. I believe a better strategy is to focus on continuously growing yourself, becoming a bigger person through continuous learning. Then, as a result, bigger and better opportunities will naturally come your way. In the chapters to come, we will see several simple and practical ways to grow and prepare yourself for doing well in your current role while getting ready for the next steps and beyond. And while you are becoming a bigger and better person, you will also learn how to create and recognize opportunities for yourself, and build the courage and strength to jump at them.

I do recommend a few general approaches. I always sought out new positions that offered the most growth. I would gravitate toward assignments that provided the greatest challenge and change that someone would allow me to take on, because to me, more change means more growth. Think of the times you experienced the greatest changes in your life. Maybe you entered a new school, moved to a new location, found a new job, a new spouse, a first child. Yes, there was a lot of upheaval, and you somehow survived. But not only did you survive, you grew a lot. Change equals growth. Gravitate to it. I know that this is really hard for most people. It is so much easier to stick with what you know and what you are good at. Stasis is comfortable and change is daunting, This is fundamental to being human. But never forget that comfort is the enemy of growth; be wary when you find yourself getting too comfortable in any situation.

I cannot emphasize enough that *the drive for change and the resulting personal growth is perhaps the most important difference between successful and unsuccessful people*, and is one of the most important themes of this book. If this is the only lesson that you take to heart, this book will have already served you well and returned to you many-fold what you paid for it.

I would not recommend anyone taking a stepping-stone job that they know they won't like, in hopes of getting another better job later. Surely everyone needs to pay their dues, but if you end up in a job you don't like, such as working for a boss or company you don't respect, how motivated and driven and passionate will you be? Not very, in my opinion. How can you succeed in a job when your heart isn't in it? It follows that unsuccessful people rarely attract new opportunities. so your stepping stone may turn out to be a setback and a dead end, after all. On the other hand, be careful to not talk yourself out of opportunities that come your way just because they are different or unanticipated. It is hard to know in advance whether you will like or dislike a job you have not yet done. If you are unsure, that's OK. A little fear and discomfort during a transition is a good sign. Again, go back to the first main principle that your growth is the key to your long term success, and if the new job offers you lots of growth, my bias would be to go for it. Don't let discomfort or fear-of-failure to dissuade you. If you don't fear the new role just a little bit, then it probably isn't really offering you a lot of growth.

We'll talk more about paying your dues, later in Chapter Four, which you'll find is not a bad thing at all.

A good career plan will generally include seeking and joining great organizations, and especially associating with capable coworkers and leaders. It is fine to embrace organizations with missions you feel good about, as this will be intrinsically motivating. *Even more important are the people you work with, as you will learn much from them; and their friendship, trust, and respect will do as much for your career as will your financial or organizational impact.* Though you will get a great deal from your colleagues, it is far more important that you focus your energies on what you can give them, and how you can help them succeed. As the great Zig Zigler told us, "You can get anything you want out of life; as long as you help others get what they want out of life." Don't look for the *quid pro quo* and don't keep score over who owes whom. Just give, give, give out of habit and have faith that they will someday return the favor. And by all means, you need to make a concerted effort to stay in touch and to continue to look for ways to add value (See Chapter Twenty-one on Networking).

Personal Experience

I first got involved in chemistry somewhat against my will. I was co-opted to take chemistry in high school by a guidance counselor, despite my objections that I had no plans to pursue science. He told me that there was no way I would be considered for college scholarships without 'Chemistry' on my transcript. Coming from humble means, I knew I would be unable to afford college without financial aid, so I acquiesced. I ended up taking the chemistry-for-dummies version, while most of my friends took the advanced course. I was fortunate to have great teacher, the late Albert Dolan, who got me interested in the subject. The "Chemistry for Dummies" version of the class was exactly what was on the standardized tests, and I scored well. I was told that chemistry majors were few and far between, and that if I wanted to get into a top college, declaring chemistry as my pre-major would help me gain admittance. This was good advice, and amazingly, I entered the Stanford Class of

'76, almost certainly the last guy to make the cut. I had a succession of great chemistry teachers in high school and college, and it wasn't until I had a really bad teacher that I knew chemistry was right for me.

It was also never my intent to go on to graduate school; I was sick of school after so many years and was anxious to get a job and join the real world and make some money. That also changed rather accidentally. After two years of organic chemistry, we finally got our first course in inorganic—the chemistry of metals and salts. I liked the course and talked to the professor—a shy diminutive guy named Henry Taube—a nice enough fellow but not a very inspiring teacher. I asked him why we had no inorganic labs; Stanford was all about organic and physical chemistry at the time. He said there were no plans for inorganic chemistry labs, but I could work in his lab doing undergraduate research if I wanted. He couldn't pay me, but he would give me credit hours toward graduation. I thought this was an interesting opportunity, so I signed up. After a couple years of undergraduate research, I observed that graduate students had it made! They got their tuition paid along with a comfortable stipend to live on, and pretty much did whatever they wanted without a lot of pressure. They were having a blast. So I decided to continue my education, after all. My mentor recommended that I go to Caltech, and I found a new, junior professor there who agreed to allow me to test some ideas I had thought of while working in Taube's group—regarding how metalloenzymes transport electrons. One postscript—Dr. Taube would win the Nobel Prize in Chemistry a few years later, no help from me, by the way. I had no idea who he was or how special an opportunity I was given at the time. As I said earlier, there is a lot of luck involved in life's choices.

It was a great year for jobs when I finished grad school; my thesis was strong; and I had lots of choices. I interviewed with GE only because a good friend of mine had gone to work there and the on-campus interview with him gave me a chance to catch up. The main reason I did the on-site interview at GE a few months later was because I was able to combine the GE interview trip with the Kodak trip—Kodak was one of the places I actually wanted to work. Of course, the GE interview worked out the best and I ended up there.

At GE, I chose to work on coal because of the freedom the assignment offered me. GE had purchased a coal company, Utah International, and they set up a research program in the hopes that the team would develop relevant expertise should it ever be needed. It wasn't. For three years, we were largely irrelevant and invisible to the company; and then Utah International was sold off to Broken Hill Propriety (BHP) in Australia. By this time, we had established a number of government—funded projects of quasi interest to GE's gas turbine, jet engine, and locomotive businesses, so I would end up staying with the coal group for a total of seven years.

In my last coal project, I was involved in injecting powdered coal in the form of a coal-water mixture into diesel engines. We observed that fuel injector wear was a big issue with coal. I happened to carpool to GE with a good friend and neighbor who had been leading a research team that was developing diamond coatings, ones that were made by treating methane gas with microwaves. Together we got the idea to coat fuel nozzles on the inside with diamond wear-resistant coatings. The idea worked fine, we got a patent together, and when he got promoted to lab director, he asked me to take over the diamond group for my first management assignment. That began my management career and moved me from coal to the science of diamonds, friction, wear, metallurgical and ceramic coatings.

As you may know, diamond is pure carbon, and GE had invented Man-Made® diamonds in 1954. Now by the 1980's, GE had a pretty nice business, GE Superabrasives, making synthetic diamonds for use in cutting stone, as bits for oil drilling, and now, diamond coatings. While I was leading the diamond R&D team, one of my projects was to try to make diamond films containing fluorine. Fluorine-doped diamond films were like a high temperature version of Teflon® so when the Air Force came to GE looking for help to develop anti-icing coatings for GE's jet engines, I tried to use this material. The idea didn't work, but the same group later got working on aircraft coatings to prevent insect debris from sticking; they remembered me and asked me to help.

At the time, GE and NASA were working as partners to make airplane surfaces so smooth, that the air could flow in laminar streamlines over the surface, rather than in micro-turbulence. If truly laminar airflow could be achieved, something of a manufacturing and design *tour de force* in its own right, fuel efficiencies would go up several per cent. One unsolved issue was insect debris—bug splatters on the wing and nacelles (the shell that houses the engines) during take-off and landing. Bumps of bug-guts on the leading edge destroy laminar flow all the way from the front to the back of the plane. My assignment was to figure out how to prevent them or get rid of them. We had a great time growing insect larvae into flies and mosquitoes and moths, and injecting them into a wind tunnel, and splattering them at three hundred miles per hour onto various surfaces all while making high speed videos. We tried dozens of coatings, and eventually, we found a special silicone rubber that did the trick. As part of the project, I got involved with the folks in our silicone rubber division, and ended up there in my next job.

Tell me how anyone could have planned that career progression—from bioinorganic chemistry to coal to diesel engines to diamonds to rubber!

After a few years working on rubber, my old carpooling friend became VP of Engineering in our light bulb division, and asked me to run one of his groups in Cleveland. I took the job, working with him while leading GE's halogen lamp engineering team. Shortly after moving to Cleveland, my pal was promoted again, this time leading R&D for GE Plastics. And a couple years later when GE made a tender offer for Honeywell, I was called back by him for the third time, this time to lead the due diligence and integration efforts for Honeywell's materials R&D.

Unfortunately, after a year of kicking the tires at Honeywell, the European Economic Union vetoed the Honeywell deal ostensibly for antitrust reasons. In reality, the EU director at the time, Mario Monti, the same guy who would become prime minister of Italy in 2012, had something of a personal vendetta with GE's CEO, Jack Welch, and that was what really killed the deal. Whatever the reason, the upshot was that I was left without a job. I had the opportunity to transfer to Mt. Vernon, Indiana to lead a GE Plastics R&D group making plastic films, but right when I was about to go, a headhunter called and instead found me my next job as VP of Research and Development at Nalco, a leading water, paper, and energy chemicals company. Two years later, Nalco was purchased in a leveraged buy-out, and I got let go. It wasn't personal—they let most everyone go at my level, all with some nice parting gifts!

We'll talk more later (see Chapter Twenty-Three) about headhunters—more properly called executive recruiters. I will tell you now that these guys are your best friends. Never turn down a call from them and even if you aren't interested in what they have to offer, and go out of your way to find them good people who are. I have made some great friends among the search people, who have universally impressed me with their capability and competence. But be aware that there is a gulf between retained search executives and those who work on a contingency.

While I was at Nalco, one of my headhunter friends asked me for names of candidates to head R&D at Cabot Microelectronics Corporation (CMC), a small supplier of advanced polishing compounds to the semiconductor industry. I gave him a few names, and remarkably, one eventually did become head of CMC's R&D—as my successor eight years later! Once I was released from Nalco, the same headhunter pushed me to interview for the job. It wasn't among the size and scope of jobs on my radar at the time, but I reluctantly agreed to interview mainly to help my headhunter friend out, who needed to present good candidates to Cabot. Plus, it was close by and it would be good practice after a couple years away from the job-hunting game. After the interview, I had my doubts about CMC, but in between the interview and offer, they brought on a new CEO, Bill Noglows. We met over a beer his first week, hit it off, and I signed on as his first hire.

Along the way, my old grad school PhD advisor had founded a company and asked me to serve on its Board of Directors. It then spun off a second company, and I moved on there as outside Director, as well. A few years later, this small-cap company was successfully sold to a major global plastics enterprise, and I gained valuable board experience.

In yet another unplanned opportunity, after the breakup of Nalco's leadership team following our sale to a private equity group, our Sales VP landed as CEO of a small, public biotech company, and he invited me to join their Board of Directors as well, where I served for four years leading their compensation committee and sitting on their audit committee.

The point is—very little of my career was planned. I didn't want science, didn't want grad school, didn't want GE, didn't want management, didn't want to leave GE, didn't want to leave Nalco, didn't want to come to Cabot. I bounced around from bio-inorganic chemistry to coal to diamonds and ceramics, to light bulbs and rubber, water, paper, energy, semiconductors! Just try to plan that out on a spreadsheet! And despite myself, I am very happy with where I am and where I've been. Not exactly how they draw it up in the books!

Was it all random? All luck? Totally stochastic? No. To begin with, opportunities don't just randomly jump into your lap. I was lucky that my guidance counselor in high school cared enough to steer me into chemistry, but I must have had enough on the ball for him make the effort on my behalf. I was lucky that my first chemistry teacher was so good, but I also worked hard in the class and developed an interest and aptitude for science. I was lucky I got into Stanford and happened to take inorganic chemistry from Dr. Taube. But I also was a good enough student to (barely) get into a great school, and eventually bold enough to take the initiative and approach Dr. Taube about what I saw as a deficiency in Stanford's curriculum. I was lucky my grad-school friend was the campus interviewer for GE but it was the strength of my thesis and the desire to broaden myself that got me the GE job. I was lucky my car-pooling neighbor was working on diamonds but we were both strong enough to combine our knowledge to solve an important problem. I was lucky he turned out to be a superstar with a meteoric career trajectory, but it was also because he thought enough of my capability and potential that he chose to bring me along not once, but three times to career-advancing opportunities.

I think we see a basic theme. Careers and opportunities come at you in many unplanned ways and directions. You need to expand yourself so that when opportunities come to you, you will be ready for them. Subsequent chapters will give you directions on how to do that. You need to make some good calls on which opportunities to go after and which ones to pass on, and this book will also help you there. You need to have the confidence to take on new assignments that are outside of your experiences, ones that you just might fail at. Along the way, frankly, you must never lose sight of the fact that it sure helps deliver the goods with each opportunity. Nobody expects you to succeed every time, and frankly, if you aren't struggling and missing your targets every now and then, you probably aren't taking enough risk. A good miss isn't a bad thing if you learn from it. Remember that some of the ball players who struck out the most include Babe Ruth and Reggie Jackson, guys who have found a home in the annals of sports and the Hall of Fame. In Chapter Twelve, we'll discuss what to do when you don't achieve your objectives and how to turn that into a positive experience and outcome for you.

It may be nice to think of one's career as a chess game where you plan several moves in advance, and then execute on them, culminating in the dream assignment. I would guess that some people actually do this, or think they do . . . *"I'll go to a Yale law school, get top grades, write on the Law Review, get a clerkship for one of the Supreme Court Justices, then take a job as Assistant District Attorney for a major metropolitan area, then work on the mayor's staff, run for State Legislature . . ."* Great. Why not? More power to you.

More realistic is to use each assignment for personal growth, enriching and broadening your knowledge and experience. With each assignment and organization, make sure to develop and maintain close interpersonal networks of people who are sharp and who are going places, and who like, trust, and respect you. Make a major effort to help them succeed. Keep your eyes and ears open for new opportunities that arise for yourself, your company, or your friends. It is important to take initiative, dive in, and execute like hell on each job and deliver the goods under the assumption that this will be your last assignment. By not thinking about your next assignment but really focusing on your current one, your reputation for success, especially with up and coming colleagues, will open many doors in unanticipated ways and places. Often your successes create successes for your bosses and coworkers, and they generally don't forget who helped them. And remember that it is the journey, not the destination, that really matters.

That, my friends, is my brief story, and I think it is fairly prototypical for a modern career path where frequent change is the norm. Your career is neither totally planned nor totally random. I hope you are comfortable with that, because like it or not, changing jobs, changing careers, changing venues—these are the cards we are dealt.

Chapter Two

Getting Started on Your First Job

I SUSPECT THAT MOST OF YOU already have your first job and probably several under your belt by now. Even so, I wouldn't skip this chapter since it is brief, and may offer insights into how to get your next job, since many of the same concepts apply.

There are a plethora of books designed to help you land your first job or any new job so this chapter will not be comprehensive, but may give you a new insight or two from a person who has hired dozens of people into their first jobs.

I will presume that you have completed some level of formal education or are about to, and now or someday soon, you will be seeking a first job or an entry to a new career somewhere. You most likely have some work experience, perhaps of modest relevance to what you are looking for now.

You ought to target certain organizations to apply for employment. Some of you are technology generalists with skills that could apply to any number of industries and segments. Others may be more specialized. In general, at this stage you ought to keep an open mind and an open funnel. By that, I mean that I wouldn't exclude too many opportunities *a priori*. Many great careers were launched from unusual origins. Plus, let's face it—in this economy, any job is going to be hard to find including even retail and waiting tables. I recently read that some McDonald's are requiring BS degrees and experience!

You might consider the public sector, teaching, the military, any number of federal, state and local government and non-governmental agency posts. Probably the largest number of positions can be found in the private, for-profit sector, and here the choices can be staggering. I mentioned briefly about my personal experience with very small start-up companies, midsize, large, multinationals, and conglomerates in any number of manufacturing and service industries. Each has its strengths and weaknesses and one variety might best match your interests and capabilities. Some planning and consideration of where you would ideally like to start is warranted. You might not find the exact starting point in your career that you desire, but if you at least know about where you want to start, you can target opportunities close to that.

Think about specific organizations that you would like to join and then find out if they have formerly posted openings. If you are graduating from college or if you are a recent alumnus, your placement center will doubtless act as a clearinghouse for campus interviews. Even if the company has no official openings, and may even be in a hiring freeze, do not consider this to be firm and final. In my experience, organizations always have openings, or will make one for an exceptional candidate. As we will see later, when looking at the job market from the hiring manager side, even if he has no openings, a good manager will always be searching for talent in the event of a sudden departure or if and when the hiring door opens a crack. He will be ready to pounce with an offer the instant he is given the go-ahead. And even if he is OK with his current team, a smart manager will relentlessly drive to upgrade, and you could just be that upgrade for his team. Don't be a pest, but touching base from time to time when there are no openings can position you for the first call when one suddenly arises.

Many companies offer intern programs, often for pre-graduates during summer breaks. By all means, aggressively seek these opportunities early during the school year. Contact HR departments in companies and industries you might be interested in and get on their lists. Use your friends, family, and faculty to identify possible internships that might be available. Don't just take an internship for the money unless you simply must to pay for college—look for the experience premium that will pay you back several-fold when you finally seek full time employment in earnest. An internship is a terrific way to get to know what life in this organization would be like, and for the organization to give you a cheap test drive, as well. And while you are an intern, don't just do your job. Network like hell. Take the opportunity to get to know all of the key business and technology leaders, ask them about the company, their roles, their careers, and what they did that made a difference. And ask for the sale, ala, "When I graduate, would you be willing to consider me as a full time employee?"

It is always better if you can find a direct "in" with the organizations you desire to join. Try to find friends, family, faculty, or recent alums who are already employed there and take the initiative to contact them. Most people are suckers for someone who is enthusiastic about joining a new company and who has done her homework. Have a clear and easily articulated vision for your immediate future and why you think this place is the best fit. Be personable and sincere and appreciative for any advice you can get from the inside, and by all means, again do not hesitate to ask, "Would you be willing to recommend me for employment at XYZ Corporation?" Then find out how.

Whoever you meet along the way deserves an immediate, hand-written thank you note, postmarked that day. You want to stand out? This will do it. Invest fifty bucks into some good, personalized stationery and write a sincere note with a specific reference to something about the person you spoke with. These days with e-mail and twitter and text messaging, a personal snail-mail note is rare and special and indicates extra effort—a tweet is just too easy and lazy. Handwritten paper notes have continued to work well for me throughout my career.

Your Resume

You will prepare a resume, no matter how little experience you have. Be one hundred per cent factual and never falsify or exaggerate anything on the resume. Lying is a career-killer and it is just plain wrong. I have seen major leadership roles dashed decades after a resume was found to include falsehoods and half-truths. One theme you will hear from me frequently is that integrity and character matter more than anything else. Regard yours as your most precious asset. If lesson one was to seek positions that offer you the most growth, and lesson two was to find great people to work with, learn from, and help, then *the big lesson three is to maintain your integrity to the highest standards*. Don't let those creeps in Washington who only look like they are successful convince you otherwise. If you can look yourself in the mirror each day and be proud of yourself, your life will be immensely more satisfying and fulfilling.

But for now, let's get back to your resume.

For your first job, nobody really expects you to have accomplished much so your resume will be sparse. Include your grade-point-average, because if you don't, people will assume it is bad. If you are a foreign national and have a visa issue, also include it. Nobody wants surprises. I would avoid "cute" resumes, unusual paper and fonts, and I would not include a picture. Just play it straight.

Any unique curricular or extracurricular activities, especially showing leadership, can be included. If you have exceptional references, by all means use them. The whole point is to stand out and get someone's attention. After that, the interviews are yours to ace. In Chapter Three, we will take a hard look at just how to effectively interview for a new post.

Since there are so many books and websites that discuss resume-writing, I won't say much about style and format. I will remind you that a resume is a selling document, not a biography or scientific paper. Think of what the reader is looking for, which is "Can she do the job? Is she the best I can find for the job? Does she have runway beyond what I am hiring for?"

I have a few pet peeves regarding resumes. To begin with, make sure it is grammatically perfect. Do whatever it takes—i.e. offer to pay your friends a dollar for each error they find—but get it right. I may cut a little slack to a foreign national who speaks English as a second language, but not much. People who are careless and sloppy on resumes will be careless and sloppy on the job. Other than the buck-per-typo, I am not convinced that professional resume services are worth what they charge. You know yourself better than anyone else. Take the time and effort to fit who-you-are into a standard template and then go from there.

You may wish to extoll your recent duties and responsibilities. I have minor use for this information, but include it if you can be brief. Truly what I am interested in is what you actually accomplished, not what your job description says you were supposed to do. Did you make your organization money or did you save them money? How, and how much? The more significant figures in your numbers, the more I will believe them. Did you lead an innovation? What was it and why was it important? A resume filled with accomplishments and impact is so rare, it will knock my socks off. Make yours like this.

Your Cover Letter

You will also include a cover letter with your resume. By all means, customize your cover letter for each application along with your resume if you think it will help garner attention. It is best to send the package to a name and title rather than "To Whom It May Concern" or "Dear hiring manager." Make sure you state explicitly what you are looking for, and try to convey genuine interest in the organization and its mission.

Do a fair amount of homework before you apply for a position, and use some of this knowledge in your cover letter. For example, "When I read about your project X in the December 3 *Post*, I thought that this could be the perfect place for me to apply for my first job out of college." Try to think of ways to illustrate your skills, character, and interests, and what you can bring to the party. But mostly, get in the habit of thinking about them and what they want, not about yourself and what you want.

Summary

There are numerous comprehensive books about getting your first job in particular, or most any job in general. I would read a few of these and draw my own conclusions. Surely they will advise you on ways to present yourself in the best possible light. I hope they will also advise you to be yourself and to be honest and sincere, because if a company hires you under false pretense, this is a prescription for disaster down the road.

Chapter Three

How to Interview

ALMOST CERTAINLY, YOU WILL BE INTERVIEWED several times throughout your life. It is also likely that you will sit on the other side of the interview table from time to time. Mastering the interview process will be critical for your landing the right job, and for hiring the right people—two of the most important processes in your career. This chapter with give you an overview of the interview process and strategies on how to handle it. There are a number of great books and articles on interviewing including *The Smart Interviewer* by Bradford D. Smart, *How Would You Move Mt Fuji* by William Poundstone, *Don't Blow the Interview* by Ralph Ferrone, and *101 Great Answers to the Toughest Interview Questions* by Ron Fry.

I am rather philosophical about the interview process. Someone once described interviewing as 'getting married after just one date.' You really want to get this right. It is not about gaming the system or pretending to be someone different from who you are in order to get the job, any more than you would pretend your way through the dating process and end up in a horrible marriage. Certainly getting an offer is one of your objectives as a candidate, but it is far better is to be yourself and to make sure that you are what they really want and conversely. Pretending to be someone you aren't, in order to get an offer, can be a disaster for all. If you do get hired, both parties will be disappointed and lots of time and money will be wasted.

On the other hand, both you as a candidate, and the hiring organization will certainly want to present yourselves in the best possible light to be appealing to the other. There is nothing wrong with that. Just be sincere and honest. One of the reasons I like "situational interviewing," which we will see in a moment, is that there are no right answers; the questions are designed to find out who you are and if there is a genuine match.

Being someone you aren't during the interview is just as bad as faking your resume. I cannot emphasize enough that, no matter what directions your career takes, make sure your character and integrity are one hundred per cent spotless. Your reputation is your most important asset and is irreplaceable. Never lie on your resume or during the interview process. Period.

Insofar as your resume and contacts helped you get the interview, let's presume that someone thinks there is a close enough match between your skills and interests and those of the organization who are considering hiring you. Now they have you in the flesh to probe whether there is real substance, character, and a cultural match.

Before we get to the actual interview, there are a few *faux pas* to avoid. By all means, make sure you show up slightly early, say five to fifteen minutes, but not much earlier. Be well-dressed, but in a manner appropriate for the organization; suit and tie for men or dress suits for women are appropriate for most businesses but be careful if this would be considered inappropriate. It is OK to ask the person who contacted you to set up the interview for advice on interview attire. By all means, be clean and neat, and avoid too much cologne or perfume. Do not wear suggestive clothing and try to hide tattoos and piercings. Do not smoke or chew gum during the interview. Do brush your teeth and use mouthwash.

Prior to the interview, make sure you do lots of homework on the organization. Read everything you can and speak to employees or suppliers or customers whom you think might have valuable insights about your target company. Try to gage what are their current issues and hot buttons. Learn about each of the organization's branches and its leaders, locations, industries, strategies, and anything in the news. Plan on spending at least eight hours doing research about the organization if not more for early-career jobs, and much more for senior positions. Joining a company is a big commitment and you really want to make sure it is right for you and them; is eight hours asking too much? I don't think so. If you can find out who your interviewers will be, by all means learn as much as you can about them as possible using public and even personal sources. They will be impressed that you did your homework.

Always treat everyone along the interview process kindly, especially receptionists and administrative assistants, and waiters and hostesses if you dine out as part of the interview. For each person that you encounter, write down his or her name and title and notes about anything you might have discussed; later that day you will hand-write a personal note to each of them thanking them for their time, and include an anecdote about something you heard during the interview. For example, to the grandmotherly administrative assistant that you chatted with while waiting for the boss, you might thank her for making travel arrangements, and comment that 'you hope her grandsons hit home runs in tomorrow's little league game' that she said she was going to attend. To the sales manager you met, you might send 'I hope you win that Hawaiian cruise sales contest you mentioned; if you hire me, maybe I can help push you over the top!' Try to say something cute like that—short and sweet. I wouldn't use the follow up courtesy-note to elaborate on some of the interview questions you think you may have messed up. The goal of the note is to show that you are thoughtful.

In addition to the substantive Q&A sessions, remember that all the social interactions, your appearance, demeanor, style, grammar, and preparation are fair game for the interviewers, and can be critical to the hiring decision. Despite the obvious emphasis on your skills and ability to answer interview questions, the entire package from the moment you arrive to the time they have read your personal thank-you notes the next day, are all components in your assessment. Do not downplay them; people generally hire people they like as much as they hire for specific skills and talent.

Now let's talk about the formal interview. A good interviewer will learn about you—what makes you tick; skills, interests, character, and runway. The desired outcome is to determine whether there is a good match between you and the organization. We take on face value that this is the only goal of the interview process—to find the best person for the job. Unfortunately, I haven't encountered very many really good interviewers, though the executive search folks are universally outstanding in this area. For them, this is their *raison d'etre* and being good at it is crucial for their very survival. Most other hiring managers sort of muddle through the process, think they get to know the candidates well enough, and often wonder why they are eventually disillusioned when faced with a separation not too far down the road.

Some interviewers make their hiring decision within the first few moments of the interview based on an emotional connection, and fail to probe for inconsistencies or for contrary views. As an interviewer, once I find things I really like about the candidate, I try very hard to look for contrary views; and conversely, if I see something I don't like about the candidate, I will probe for more positive alternative views.

Even if the interviewers are mediocre, since you will be thoroughly prepared for the process, you can actually help steer them to learn what they need to know about you. Let's see how.

Most interviewers want to hear about you through stories and vignettes that illustrate how you behaved in the past as a predictor of the future. Some interviews are speculative—"If you were in such-and-such situation, how would you behave?" These are not quite as valuable as "*How-did-*

you-behave" questions, but you can turn this around if you refer to an actual situation that illustrated your performance and character. For example, if an interviewer poses a hypothetical situation to evaluate your courage, you might turn it around by directly saying something like, "I think you are probing whether I would behave courageously in this situation. I'm pretty sure I would, having faced something similar in the past. Here is what actually happened . . ." See what I mean?

To prepare for your interview, make an inventory of your "stories." Think of specific incidents, instances, actions and activities of your past that you can use to illustrate your substance. Do not limit yourself to professional situations—your whole life should make up your palette including school, sports, community, religious, family, travel. Si Ramo, the "R" in TRW and founder of the defense-electronics industry, got his first job at GE in 1929 during the heart of the Great Depression because his resume indicated that he was a virtuoso on the violin; his hiring manager just happened to direct the Schenectady Symphony Orchestra and he was one violinist short at the time. The rest is history.

In the pages that follow, I will list some of my favorite, and toughest interview questions that I have heard, read, or asked over the years. I am always on the lookout for great questions to ask; and to those of you reading this book who might have given me one of these, I thank you without fair attribution. Sorry, I wrote down the questions but not the source.

Please practice answering each question as if you were asked it during the interview, and use the questions to inventory illustrative stories.

Let's take a look at the Spiro 100 Favorite Interview Questions:

1. Walk me through the various education and career decisions that you have made. Tell me why you made the choices you did.
2. Tell me about yourself.
3. What are your strengths?

4. What is your greatest weakness (ugh!)
5. How do you organize your time? (A red flag is just one day at a time *vs.* a long range plan with milestones).
6. What aspects of your work are most critical?
7. What aspects of your work (or tasks) do you enjoy the most?
8. What aspects of your work (or tasks) do you enjoy the least?
9. How many hours does it take you to do your job each week?
10. What are time wasters for you?
11. How important is communication in your work?
12. In what ways do you communicate with colleagues, and which is most effective?
13. What decisions are most difficult for you to make?
14. What are the three hardest decisions you ever made?
15. What is the best decision you ever made?
16. If could go back in time and undo one decision you made, what would it be?
17. Give an example of a time you had to make a quick decision with incomplete information.
18. What are you curious about?
19. What have you gotten in trouble over?
20. What have you done that made you laugh at yourself?
21. What makes you so angry your blood boils?
22. When was the last time you lost your temper? What caused it and how did you respond? What was the eventual outcome of the situation?
23. How have you grown this last year?
24. When were you disappointed this last year?
25. What has been your biggest disappointment ever? How did you deal with it?
26. What are the three greatest risks you have taken in life?
27. Where have you travelled?
28. What is your favorite place?
29. What do you do if you are alone at night in a strange town?

30. If you had to choose between a week's vacation at someplace new that you think would be great or someplace you have already been and knew it was great, which would you choose?

31. Do you consider yourself successful?

32. Do you consider yourself lucky?

33. Give me an example of a time you showed great courage.

34. Would you consider yourself more competitive or cooperative?

35. Which drives you more—the joy of winning or the fear of losing?

36. Think about your best boss; what characteristics did you like most?

37. Did your boss and you have lots of interaction or did she leave you alone?

38. What supervisory style do you prefer?

39. Tell me about your worst boss. (A red flag is identifying the person by name instead of generically).

40. Give me an example of a time your manager opposed a suggestion of yours; what happened?

41. When you leave, what will your boss miss least (most) about you?

42. If I were to contact your last boss (or other people the candidate might have mentioned) and ask him about you, what would he say?

43. Would it be all right if I contacted your last boss or previous supervisors?

44. How did your last boss influence your decision to look at this job?

45. Why are you leaving/why did you leave your current job?

46. Why did you leave each of your previous jobs?

47. Describe a situation where you encountered cultural barriers that hindered your effectiveness.

48. Describe a time you had to decide whether to launch a product or service before you knew whether it was fully capable.

49. What is integrity to you? Give me an example of a time your integrity was put to the test.

50. For managers, "What is your turnover rate?"

51. What were the reasons your employees left?
52. Tell me about your favorite employee or favorite coworker.
53. Tell me about your least favorite employee or least favorite coworker.
54. What do your people (or colleagues) think of you? How do you know?
55. What is the biggest misperception others have about you?
56. Describe yourself in three words.
57. Who are your heroes?
58. If you could spend an evening with anyone from the past or present, who would you choose?
59. What is/are your favorite web sites? Books? Movies?
60. What kinds of books do you read?
61. What are you reading now?
62. How are you unique?
63. Give me a list of "I likes" both personal and professional.
64. What is your favorite interview question?
65. What would you do if you won the lottery?
66. What is the highest pressure situation you worked under?
67. What do you procrastinate over?
68. What are some innovations that you introduced in a project/class/organization? How successful were you in implementing?
69. Give me an example of a time when you had strong convictions about a course of action but were subsequently convinced to use another approach.
70. In trying to achieve goals we often run into obstacles. Tell me about an important goal, an obstacle you ran into, and what you did to overcome the obstacle.
71. Give me an example of a time when you had to organize or coordinate a group toward achieving some goal.
72. Tell me about a time when you had to plan and organize something. What did you do? How did you choose the resources you needed? What did you do to make sure your plan was working?

73. Describe a situation in which you had to work with a person you disliked.
74. Give me an example of a time when you had to sell a tough idea.
75. Tell me about a time when you personally influenced someone to take action he/she initially resisted?
76. Give me an example of a time when you were particularly effective in communication with another person.
77. Give me an example of a time when you were particularly ineffective in communication with another person.
78. Describe a time when you exceeded expectations.
79. Describe a time where you fell short of expectations.
80. What school activities did you participate in and why?
81. Describe some of the leadership roles (teams, clubs, fraternal, etc.) you have taken and your responsibilities.
82. Are your career plans more oriented toward your specialty or toward managerial positions? Why?
83. Tell me about a situation that did not work out in your life. How did you deal with it?
84. Give an example of an important goal you established. How did you reach it?
85. What is the most creative work related project you carried out?
86. What about our organization makes you want to work here?
87. What do you think you could contribute to this position?
88. How are you unique? What differentiates you from others?
89. What do you want your legacy to be?
90. What was the best job you ever had . . . worst job you ever had?
91. Give me an example of a time when your schedule was upset by unforeseen circumstances.
92. Describe an unpopular decision you made.
93. How was your education funded?
94. Describe a time when you discovered, or were told, about a deficiency in your job performance; what did you do?
95. What frustrates you about your current job?

96. What has been your most satisfying work outcome? Least satisfying/most disappointing?

97. How do you motivate others? Give an example of someone you successfully motivated . . . and an example of an unsuccessful attempt to motivate someone.

98. How many gas stations are there in America? How many hospitals are there in America? What is the mass of Mt. McKinley, the volume of water on the planet? (Look for analytical thinking, multiple ways of estimating, do not allow people to guess). Look for reasoning processes.

99. Is there any question I haven't asked you that I should? Or, "We only have about 5 minutes left in the interview; is there anything I should know about you?"

100. What questions do you have for me?

For the moment, let's just use these questions for practice. Take each question and think about how you might answer it. If there is a request for a specific example, say question 40 looking for a time a manager opposed a suggestion of yours, try to think of not just one example, but as many as you can think of. For some students of the interview process who have learned how to game the common questions, asking for a second or third example of illustrative behaviors cuts them cold. This is one of the reasons I ask people for the three greatest risks they have taken, rather than just one. While practicing for question 74 on selling a tough idea, remember that it is fair game for the interviewer to say "Wow that was interesting—give me another example." Moreover, as you prepare your personal vignettes that illustrate behaviors, it is nice to have many to draw from as they often illustrate several aspects of character and capability.

It is likely that, depending on the length and drama in your life, you will have perhaps a half dozen major stories and perhaps ten or twenty illustrative vignettes. Think about them carefully and what they say about you to an interviewer.

One key to a successful interview is to think about what the interviewer is really trying to get at when she asks you a question. For example, when she asks you, "Where have you travelled?" she probably isn't terribly interested in a travelogue about your recent vacation. So don't just say "I spent a semester in Taiwan, took two European vacations, and went to Aruba every Easter with my parents." You may have answered the question factually, but think of the meta-question and reach into your stories. Add something to the effect that "I love travelling and adventure; I'm always looking to grow and expand my experiences. I love to experience different cultures and I truly embrace diversity. One of the most important lessons I learned by living in both in Asia and America is that deep down, people are all generally the same, but that there is no single best way to live and do things."

Let's try another. Say question 95, "What frustrates you about your current job?" It is never a good idea to say anything bad about your current employer, boss or coworkers, so be delicate here. It is perfectly fine to say things like, "Things have been going well there, but I would like to see if I could accelerate my personal growth and career by joining a new organization with different products, business processes, and people." And you can expand, as in "I found that working under government contracts was somewhat encumbering. I understand the need for rules and regulations to ensure that the taxpayers are getting everything they are paying for, but I found that there was a lot of red-tape that didn't really add value toward achieving the desired results, and achieving the performance targets is what really gets me enthused about the job, not filling out reports."

Frankly, I really doubt you will get as hard an interview as I give with these questions, at least early on in your career. Some of the big business consulting and investment banks, as well as Google and Microsoft may actually be tougher. And later in your career as you encounter executive placement (Chapter Twenty-three), you can count on strong, probing questions. But for now, go through each of these questions and practice answering, reaching deep into your well of experiences and writing down all your stories to draw upon. I hope you will find it to be valuable introspection. One candidate, faced with my "three greatest risks" question, kind of shook his head and said, "You know, I probably am not taking enough risk in my life as I think about it." You just might find some valuable insights, yourself. It is always better to face them before you are on the spot in the interview.

And just a few general tips on the interview, itself. It is good to smile, be enthusiastic and optimistic. Eye contact is good but staring is bad; make sure you know the difference. A firm handshake is good, including eye contact, but there is no need for bone-crushing. If you are nervous and shaky, it is OK to acknowledge the fact as an ice-breaker by saying something like, "I appreciate the opportunity to meet with you. I think this is such a great opportunity for me that I hope you'll bear with me as I am a little more nervous than usual." If you are prone to cold, sweaty palms, a little talc splash in the car beforehand can take care of that, but be careful to not spill it on your black suit! Unless you are applying for the clergy, the two handed handshake is a little over-the-top. And despite the obvious tension, try to relax as much as possible, take a deep breath, and consider the person across the table as a friend and ally, not an adversary.

When facing tough questions, it is perfectly fine to take a moment to think of your answers. Candidates with snap answers to tough questions make me suspicious. Do not dodge any questions, even if you can't think of a great example. It is always better to dig into your well of stories and come up with something close. Politicians are great at turning around tough questions and giving answers that may not exactly be what the questioner wanted, but which show them in a good light.

Try to keep in mind that, while it may seem as if the interviewer is holding all the cards. In fact, if you are a good candidate, and let's certainly hope you are, the hiring company has a great deal to gain by adding you to the payroll. Regard your conversations as between equals—two parties simply trying to determine if there is a match. Be confident. Make the interview conversational. Keep your answers brief enough that they would seem to be normal conversational discourse. You wouldn't speak for five minutes straight to your friends; don't run on in the interview either. You don't want ridiculously short answers, either. Be concise; make your point; and ensure that the questioner learned what he needed to before moving on. It is perfectly OK to ask if you satisfactorily answered his query.

Often you will be asked if you have questions, i.e. number 100 on my list. You really should use this opportunity effectively. Surely there are a myriad of things you are curious about. What does the interviewer think about the company—its strengths and weaknesses and future? How are people measured and rewarded and recognized? What is morale and is it trending up or down? Frankly, if you can't think of dozens of important questions, you probably ought to not be interviewing for the job. And remember that your true strength is often better revealed by the quality of the questions you ask, not the answers you give.

You As Interviewer

Eventually, you will be the hiring manager who is sitting on the other side of the table asking questions. The quality of people you hire is perhaps the single greatest determinant of your future success once you have moved into a leadership role. Take this process very seriously. You should certainly use many of the Spiro 100; I would often read through a candidate's resume and then print out my list and check off or highlight the ones that I thought would be revealing, especially in areas where I might have doubts. For example, if I were concerned with creativity, I might ask question 68 about innovations the candidate might have introduced. Or if leadership was a concern, I might ask #72 about organizing a group. Normally I will have several issues that I want to address, and will script a series of questions. Usually the candidate will tell me a story that addresses several of my issues, and I can skip around.

In general, you should be a very good listener when you are interviewing, focusing more on the answers you are hearing rather than on the questions you will ask next. By having the full set of questions printed out and at your fingertips with key ones highlighted, you can truly focus on the candidate. In addition, having a script makes you as an interviewer look professional, serious, and prepared.

As a good rule of thumb, while mainly conversational, you as the interviewer do not want to be speaking very much. After all, while you are talking, you are not learning about the candidate, and that is the objective. If you can get them speaking north of eighty per cent of the time, you are doing well.

I will try to start the interview by putting the candidate at ease, asking some social questions such as, "Any trouble finding the place?" or "How was the flight?" I then explain the process, how I will be asking a series of questions that have no right or wrong answers, but are just designed to see if we have a good match between the candidate and the company, and that they should be as open and honest and complete as possible. And I will let them know that I will leave some time at the end for them to ask me questions. I will then go into my scripted questions.

I will actually make notes on the scripted questions as a reminder of their answers. If you are interviewing several candidates, these notes can be invaluable as several candidates will tend to blur into one another.

I will usually start with question 1, "Walk me through the various education and career steps." This gets a rather lengthy answer. I look for the substance of their answers—good choices, good thoughtful decisions—and also how concisely they answer. Most individuals provide a chronological blow-by-blow account of their lives, and this is OK. Far better are those who offer a theme. For example, a great theme might be, "At each step along my career, I sought to surround myself with better and better people and institutions that caused me to grow and stretch just to fit in," or "I have fought to avoid being pigeonholed into narrow areas, and always sought educational and career steps that provide greater breadth of experience, as I feel that the most significant issues require examination from many diverse viewpoints." If you get this, fasten your seatbelts—you probably have an exceptional individual before you.

Usually, in the course of her chronology, the candidate will highlight certain situations that will match up with your questions of interest. For example, if they say something like, "I worked for XYZ Corp for three years, then moved on to ABC Corp . . .", you might follow with, "What influenced your decision to leave XYZ Corp?" or "If I contacted your boss at XYZ Corp, what would he say about you?" Sometimes a good place to start with experienced candidates is, "What was your favorite position . . . least favorite position?" You will see that the conversation flows quite naturally into your questions.

If you have an area of concern and the candidate is rambling too long in an area that you have found to be quite satisfactory, it is ok to just sort of put up your hand and say something like, "Jerry—I have a good idea about this particular situation, but because we are so limited in time, I really need to ask you about—such-and-such." It is a little bit rude and abrupt, but it can be done gracefully with practice and most candidates are understanding.

Occasionally your candidate will get stuck and not have an answer to your question. Wait them out! Don't let them off the hook too easily. If they say, "I can't think of a good example." say something like "Try harder . . . I'm sure there is something similar in your past." Or just wait quietly, looking at your list of questions, make some notes, and let

31

them fill the embarrassing void of silence. Once they get the idea that you will not settle for nothing, they will reach for answers that will often be most revealing.

I really like to wind the interview down with "We only have a few minutes left; is there anything I haven't asked that I should know about you?" Occasionally you will learn about some serious new issues at this point such as "I am three months pregnant" or "I have a child with special needs" or "I plan to commute from Chicago every week—will you pay for that?" And then finally open the floor up to their questions.

Now before you go off as an interviewer and pull out this list, I should warn you that many organizations do not espouse this behavioral interview process, so you will need to clear these questions with your human resource or legal departments. I don't personally believe that they are illegal as long as the questions help illustrate behaviors and character necessary to succeed on the job, but there are others who may differ, so be careful. There are many strict laws on the interviewing process to protect candidates from employer—bias based on race, creed, color, age, gender and disabilities.

I will say that, having used these questions for decades now, I have never been sued or even reprimanded, though at least a half dozen times I unwittingly brought candidates to tears when I touched a nerve. For example, I asked one woman "Who are your heroes?" and she burst out and wept uncontrollably. It was not long after September 11, 2001 and she thought of the firefighters in NY City. Insofar as I happen to be a volunteer firefighter, I thought her response was perfectly appropriate and eventually offered her the job. But she was mortified by her outburst. Be aware that asking candidates about problem coworkers or bosses, or particular frustrations and disappointments, can be touchy.

On occasion, the candidate will just dazzle you and you will know on-the-spot that you have found an ideal match. At this point, don't be shy about letting her know how you feel. I will go into selling mode, and let her know that I will be strongly recommending that we make an offer. Candidates really like this sort of immediate positive feedback, and your quickly wanting to hire them will factor heavily in their decision to eventually take your offer. If they are that good, you can expect that they will have several offers to choose from, and your instant recognition and appreciation of their talent and your decisive offer will stand well with them. Everyone likes to feel that they are truly wanted and needed, and most people also like decisive supervisors. This is especially effective if another competing hiring organization is slow, deliberate, and political.

More often than not, you may like the candidate, but will have some doubts. After all, people are complex and it is unlikely that anyone will exactly meet all of your needs. Deciding whether to take a chance or not is a very difficult decision, and is best made by getting inputs from several sources including your entire interviewer panel, references, and through a process of comparison with several other candidates. You can wait for a very long time before you find perfection, so you will have to accept some shortcomings. Are there things that you don't like but can live with? Does the person have the right skills, character, values? Would she have good chemistry with you and the rest of the team? Does he have runway and leadership potential? How coachable is she? In the end, you will be successful based on the quality of your hires and you will be judged based on the team you put together. Important stuff indeed.

Assuming you will be interviewing several candidates, you owe it to each of them to let them know the process and timetable, and to stay in touch afterward. Companies who never give feedback or responses can unnecessarily cause hard feelings down the road. It is perfectly OK to say that you have filled the position and that you appreciated their coming for an interview, and even that you will keep them in mind for positions that may be a better fit in the future—if that is the truth. But silence is unacceptable and rude.

You may see a really good person on paper who answers most of the questions well, but you hear a niggling alarm in the back of your mind saying, "Stay away." I would tend to listen to this voice, especially as you gain experience. There may be subtle cues on character or personality or substance or style that you can't quite discern, but you can't get away from them, either. You need to trust your instincts. On the other hand, be very careful if the only people you feel good about meet a particular mold, especially if this mold is who you are or who you'd like to be. People are often unaware that they have this built-in bias. Maybe they like to hire jocks, glamour, or jolly chubby types. Maybe they have soft spots for smooth talkers or wry humor or even a British accent. If you see homogeneity creeping into your organization, force yourself to consider otherwise. Be careful to not hire people just like yourself; people who think and behave differently are the keys to you and your organization's growth. The last thing you need is a bunch of people who think and behave just like you.

I will also say that candidates who survive a tough interview and still get the offer tend to feel really good about the company that is hiring them. If the interview was easy and less discriminating, a truly strong candidate might feel disappointed that he wasn't truly given an opportunity to test his mettle and distinguish himself from the run-of-the-mill candidate. Do not feel like a hard interview will scare good future employees away. I have had so many candidates walk out shaking their head saying "I have never had an interview like that," and yet they know in their hearts that, in just an hour, I got to know them better than most of their close friends and family who have spent hours and hours with them. And really, isn't that the objective of an interview—to know the candidate well enough to make a great hiring decision? I think so.

Occasionally, you will find make an offer to a candidate who turns you down for an alternative post. That is to be expected. Do not lose track of this person. Maybe you came in second on their list. According to Harvey Mackey in *Swim With The Sharks and Don't Get Eaten Alive,* being number two is OK because sooner or later, number one is going to mess up. If you stay in touch and from time to time, ask about how things are going, and even offer help with their business or career, you may just catch them when they want to leave their current job, and you will be the first place they come to. I have hired many outstanding candidates who turned me down the first time. So don't take "no" as a permanent answer, just a temporary one that you can remedy.

And finally, practice and experience are great teachers, no matter whether you are the interviewer or interviewee. The longer you live, the more jobs and experiences you accumulate, and the more people you will have interviewed and hired with a mixed bag of results. In time you will have better stories and vignettes and you will better-recognize substantive stories and vignettes in others. You will become stronger and more comfortable in the overall process, and hire terrific employees for your organization and obtain better jobs for yourself.

Chapter Four

Some General Advice on Careers

My EMPLOYEES OFTEN CAME TO ME seeking career advice. I was always delighted to discuss career strategies with them, both confidentially and candidly. Like you, most of us are interested in understanding our career trajectory and are hoping for more, or at least, a change—something different from time to time. We hunger for growth and challenge. Many people are ambitious, seeking higher status, prestige, visibility, compensation, impact, and control—whatever floats your boat. For most, stasis is not an option, even if we are good at our jobs and like them. I like chocolate cake and lobster, but to have cake and lobster every day would get old pretty quickly.

It is extremely important that people feel that their boss or at least someone at higher levels is on their side, cares about them and their future. For those of you who are supervisors and mentors or hope to become one someday, this is very important and one of the main ways that employees stay engaged in their organization—by feeling that they have a future and that someone besides their mother cares about it. Do not take these discussions lightly. Good employees will quit over this.

If you happen to have a boss who is unethical, unscrupulous, controlling or completely self-centered, you seriously need to consider an exit strategy. There are few things more difficult on the job than feeling at odds with your supervisor. It is fine to have disagreements, but overall you both need to be pulling in the same direction and you surely need his help and advocacy. A selfish boss who hordes his talent can do a lot of damage to your career. Similarly, a supervisor who is threatened by outstanding employees is also someone to avoid. Far better is a supervisor who is grooming you to be her successor or for another advanced position elsewhere in the organization.

Recalling the last chapter on interviewing, a good question to ask your potential supervisor is about her record in advancing her employees. If she comes up blank or gives you vague answers that give you an uneasy feeling, you might take that into account before accepting an offer.

I had a most remarkable boss who actively helped me get promoted within, and even outside the company. He said, "If I can't give you the best job in the world for you, then I want to make sure I help you find it somewhere else." Wow was I lucky!

A Few Basic Principles

Let's start with the most important message: *You are responsible for your own career, and nobody else is.* There may be others who can help or hinder or advise you along the way, but only you are the one driving the bus. Yes you should certainly seek out people in your sphere who care about their employees and coworkers; you should look for good sounding boards and mentors and managers. Frankly, it is the rare supervisor who puts his employees' needs above his own. More likely, there is a tacit *quid pro quo—"You do a good job for me and help me get promoted and I will take care of you."* It is a system that kind-of works, though to me, it has shortcomings. I tend to look at all my constituents—managers, coworkers, employees, customers, suppliers, friends and family and community—and take them all into account when making career decisions. For example, if I had an employee who was absolutely vital for a project that was critical to my company and customer, I might seek to postpone his pending career move, but not always.

I understand that many of you are really concerned and confused about how to manage your own careers. Maybe you got your first job through personal or campus contacts, or through an internet search. But after that, you then just left your career to drift about and manage itself. Perhaps you just waited until someone tapped you on the shoulder and promoted you, or maybe a friend or colleague came along with a better opportunity and you took it. Basically, you took a passive approach toward managing your career. As we saw in Chapter One, life is stochastic—filled with random events and choices. This is not something anybody can control. But that doesn't mean you shouldn't still take an active role in your career. Milton was right when he wrote, "They also serve who stand and wait," but who has the patience for that?

I think "active" is the right word for career management, and "aggressive" is going a bit too far. People who wear their ambition on their sleeves are a little scary and off-putting to me. Do you know people like this? Haven't they made you just a bit uncomfortable? It is as if they don't really care about their current assignment; only to the extent that they can use it, almost exploit it, if it helps them to move along. I tend to distrust people like that. They are generally short-term in their focus in a drive to achieve results and recognition (for themselves) but with little regard for the stewardship of their organization. Hey I can always make incredible numbers just by starving research, shortcutting design, reducing preventative maintenance, eliminating new business development. I just can't make it through a couple business cycles.

Doing a Good Job Is Not Enough

A couple myths on career advancement you need to get out of your head. First, it is not enough to simply do a good job in your current assignment. That is what is expected of you and is not the main basis for promotion. It's what you do after your job is completed that gets you noticed.

Nor is simple longevity in your current position the criterion for promotion. Just because you have been in your job for two, three, five, or x years does not mean you are promotable. Yes you need to do a good job in your current assignment in order to be considered for promotion, but you need to go beyond. You are far more likely to be promoted if you have invested in, and grown yourself through reading and formal education, as well as through broadened experience. Perhaps you have found opportunities to expand your current assignment. Look for ways to make your role bigger and have bigger impact. Take initiative. Volunteer for special assignments and build your leadership resume through charitable, community, or professional organizations outside of the company. Show that you can do the next job by doing parts of it before you are promoted. If the next job involves creating strategies, try identifying new strategic opportunities and sharing them privately with the incumbent. Be subtle. Knock on the door and say something like, "If you have a minute, I was thinking about our current growth strategy, and was wondering if we ever considered X, Y, or Z as well." She might just adopt some of your suggestions, and you might learn some reasons why they won't work, but when done in the proper spirit, stepping out of your role can be a way to get visibility.

Assuming you have been in your job maybe a bit too long for your tastes, and feel that you are doing a good job, you may be wondering what is next and why your career isn't moving along faster. You may see others moving ahead of you at a much faster pace, and wonder how you get on that bandwagon. Of course, you must consider the possibility that the boss doesn't think that you are doing as well as you think. This is often the case for a couple of reasons. There could be a mismatch in expectations between you and your supervisor. Or there could be misperceptions. Most organizations these days run pretty lean and a supervisor may have ten or more direct reports, none of whom she knows terribly well. It is important that you speak candidly with her to make sure you each are in agreement on the expectations, and that she fully understands your contributions and accomplishments. It is a delicate discussion to try to ensure you get credit without appearing self-centered and conceited. But usually candor works best. Start out with something like, "I am concerned that my career seems to be stalling. I feel that I show up for work every day with a great attitude and high energy, always trying to do my very best. I have been working hard, and I feel that have consistently met or exceeded your expectations. If I'm wrong about this, I sincerely need your honest feedback so that I can jumpstart my career again." Do your homework. Be prepared to discuss specifics. And use hard numbers and hopefully, big numbers. Think about the things that she is measured on, and ask yourself how you contributed to that. This can be a ticklish conversation, and once you have invited candor, do not be defensive even if you hear things you disagree with. This is not the time to offer rebuttals. Be thankful that you are getting the truth and then deal with it.

It is rare that there is a specific promotion timetable for you or your organization. Sure, universities often make tenure decisions according to the clock, and perhaps there is a promotion timetable in the military. There may be unwritten rules that you can't move before so many years, or that you must move after so many years or you are considered stale—yesterday's news. I doubt anyone will share these insights directly with you so you will need to make your own observations and draw your own conclusions. When I was with GE, for example, I observed that it was extremely rare for new company officers to be over the age of 40. Since I was getting close to 50, I knew it that if I wanted to be a VP, it would be at another company.

Don't necessarily get discouraged if you aren't moving terribly quickly, especially if you like your current assignment. GE's Dr. Walt Robb spent many years doing research before moving into management—far more years in the lab than most people who were earmarked for management from the get-go. But Walt zoomed up the ladder and in just a few years moved to GM, then VP, then Senior VP first running GE Silicones, then all of GE Medical Systems, and finally becoming CTO for the entire company. Sometimes late bloomers end the race ahead.

In some organizations, you may be stuck reporting to a blocker— someone who is career-limited or terminal in their current assignment. For example, when I became VP of Research at Cabot Microelectronics, it was pretty likely that at age fifty, I was going to be in this role for a long time. Where else was I going to go? My tenure turned out to be eight years. Anybody reporting to me who wanted to become VP of R&D would have had to wait a long time or leave. And while I was more than delighted to help people take roles outside of R&D or to expand their roles within R&D, often you may be stuck behind a blocker who is perfectly happy with you staying in your role indefinitely, especially if you do it well. Here is where visibility outside of your immediate organization—cross functional visibility or external visibility—is vital or you will indeed be stuck.

I often hear lower level employees criticize their supervisors taking credit for what is really their work. Be careful here. Yes, a great supervisor will go out of their way to ensure their employees get the lion's share of credit when the organization succeeds, while they take the heat when it struggles. But this is a rare individual, requiring a supervisor who has a great deal of character and self-confidence. Frankly, your supervisor is probably a lot more scared about losing his job than you are, and for good reason. On the other hand, while you may feel you are not getting the visibility and recognition that you deserve, everyone up the line knows that the supervisor is not the person actually doing the work, and that there is someone at the rank-and-file level who is. And usually, they know who is really pulling his weight. Do a great job and be confident that someone, actually most everyone, knows that it is you.

If you are a hog for recognition, it actually makes you look small and insecure and conversely.

Where to Go and How to Get There

In Robert Cooper's book *The Other 90%*, referring to the common belief that we only achieve ten percent of our full potential, he describes a fascinating study begun in 1960 at a major graduate school of business. In the study, 1550 incoming students were asked about their motivations for entering business school. The vast majority—over 1250—said their goal was to use business school as a means to making a lot of money for themselves. In contrast, 255 of the students said they wished to learn of business processes that would help them become more effective in solving problems and addressing issues they really cared about.

I would guess the survey would not be much different if taken today.

Twenty years later, the students were surveyed to see who had become millionaires. Indeed a hundred and one had achieved this milestone. What is remarkable to me is that—of the hundred and one millionaires—fully one hundred had come from the group who sought to address issues they really cared about, while only one of the those whose goal was to become rich had actually become rich. Wow! As we'll see in Chapter Nine, we need to be extremely thoughtful in setting the right goals for ourselves.

So one key to life, it would seem, is *to find something you really care about and to pursue it*—the other stuff will follow.

Cooper goes on to discuss how you find these things and gives some good suggestions. Other good sources for you to consider include Marcus Buckingham's book, *Discover Your Strengths*, and Po Bronson's book, *What Should I Do With My Life?*

As you go about finding yourself, no matter what stage in your career, try to take particular notice of the things you do that come easily to you at first, or where you found yourself totally absorbed without any awareness of time. These are good places to start. For me, a few things come to mind.

Growing up in the Midwest in the 1950's and 60's, I did all the usual kid stuff—Boy Scouts, Little League, Pop Warner. When I was eight years old, my best friend and I found some old tennis rackets and went over to some courts at a nearby school and tried our hand. We instantly loved the game and despite spending most of our short lives with more common sports like baseball, football, and basketball, we walked off the courts that day both saying it was already our favorite. It was love at first sight. At that time, tennis was not at all popular like it is today, and we were fortunate to get in, and get good early, before the masses discovered the game. We both joined Junior Davis Cup, and played varsity tennis through high school where we won several championships and got good enough to play college tennis if we chose. Sadly for me, my college team was made up of future world champions like Roscoe Tanner and John McEnroe and I could barely scratch a few points off of the likes of them. But it was that instant feeling of passion for the game that struck me and is still with me today.

Now flash forward to graduate school, where I shared an apartment with a crazy Indian physicist from Bombay who felt it was his mission to invite every new Caltech Indian, Sri Lankan, and Pakistani to our place, sometimes for weeks at a time as they attempted to find places of their own. One of these guys showed me the fine art and science of counting cards in blackjack, and arranged for me to take a free junket to Las Vegas. My first trip was a crashing success and I was given a VIP card by the casinos based on how much I was betting. This entitled me to free air, food, lodging in Vegas anytime I wanted, which turned out to be once a month or so. Now recognize that counting cards is an exhausting mental exercise, and after an hour or so, you feel like you have taken a final exam in P-Chem. So one day, to take a break from blackjack, I sat down at the Circus-Circus casino's ten-cent ante seven-card stud poker table. When I finally looked down at my watch, it was about ten hours later; I had won a lot more money than I was winning at blackjack; and had more fun than ever. I was in a state of "flow" and I knew poker was the game for me. It still is, thirty-five years later, even though the stakes are a lot bigger.

Science was not so quick or easy for me. As I mentioned in Chapter One, I reluctantly took my first chemistry class only because my guidance counselor insisted that I could not get a scholarship without it. I had an excellent teacher and developed an early aptitude for science, but I was suspicious that it was only because I had a good teacher. I took the next chemistry class at a local community college also with an excellent teacher. I decided to major in chemistry, but still had my doubts. My first three professors at Stanford were all fantastic—Hans Christian Anderson (no joke!), John Braumann, and Gene Van Tamelen—all big names in science. It wasn't until my fourth class in organics by a not-so-good teacher that I decided that this was the career for me.

The last point I want to make from Cooper's book is that you may never find out what gets your juices flowing if you don't continually try new things. I still believe there is a lot more out there for me to try, and lots more growth left. People are like trees—the moment we stop growing, we start to die. So many famous and historical figures didn't find themselves until late in life including Franz Kafka, Dave Thomas, Tom Clancy, Grandma Moses and countless others.

One of the most important factors I look for in candidates and employees is enthusiasm. Given that people are successful when they are enthused about something, this attitude and energy and passion can often overcome a shortage of innate talent and ability. What are you enthusiastic about? What can you do that is seemingly timeless and effortless? Where have you lost yourself? And what new things have you tried lately to make sure you keep turning over new rocks that might be the foundation for your growth and future?

To help determine your dream job, let me recommend again Po Brunson's book *What Should I Do With My Life?* This is an insightful collection for people who are just starting out their careers or who are getting wanderlust. The key is to really think things through and to be honest with yourself. For us innovation professionals, we have an aptitude for science, quantitative thinking, discovery and innovation, a desire to understand how things work and a drive to make them better. There are plenty of great careers and lives to be had based on those interests and abilities. But there are also many more things to consider. As you move up the ladder, people become more important, influencing others, selling, finances, politics, customer relations, government and public exposure. If you loathe public speaking, you probably don't want to become a professor or CEO. If you find managing money boring, stay clear of finance. If you are an introvert, you probably don't want to be leading people.

Once you have some thoughts about who you are and dreams about where you would like to end up, work backwards from that position and see what kinds of roles and responsibilities will lead to that job.

As an example, suppose you want to someday become a university president. This is a very prestigious and lucrative position. Probably you will need to be a dean first, a department chair prior to that, and have been a full, associate, and assistant professor before that. If you look at the timelines involved, including the time to get a BS and PhD and postdoc, you are looking at thirty-plus years of planning and hard work. Still interested?

How about wanting to become CEO of a public corporation? Usually these guys begin their careers as individual contributors in their twenties starting among a breadth of assignments including engineering, production, sales, marketing, finance, law, even the military. CEO's come from many different disciplines and engineering and science are excellent foundations. Come to think of it, all of my CEO's had technical roots. Usually they are also blessed with high "EQ" or emotional intelligence, a phrase coined by Daniel Goleman. Take a look at his book, *Working With Emotional Intelligence*, to see if you have what it takes. Maybe our CEO-hopefuls had some early success and caught a mentor's or champion's eye who saw the raw potential for leadership. These "high-potentials" are typically promoted rapidly and rotated through a range of assignments in different organizations, functions, regions, disciplines, and industries so they will get broadened and seasoned. Almost always, there is significant commercial responsibility along the way, culminating in one or more profit-and-loss-responsible general management positions. At this stage, they probably need to do something bold and visionary with their roles—caretaker leaders generally do not get recognized or promoted. After all, leadership is really about change and how to drive it in an organization. I have seen a lot of folks who wanted to become CEO's who never got there because they stayed beneath the radar when critical leadership was needed and they failed to step up.

If you are later in your career and missed the early fast-track to the corner office, you seriously ought to consider going to a much smaller company including one you start yourself. I recognize that comes with some risk and probably significant financing/fund raising requirements. Still, if your own little company grows, you are already in the top spot. And perhaps you might get successfully purchased by a bigger firm. Then the chances of you moving to a big job at a bigger company are much improved.

In general, unless you have the relevant experience, it is hard to move into a very different position from where you currently are. You are better staying in a company and industry in which you are well-known where there is some possibility that someone in your company will take a chance on you stretching into a new role. In my last company, we promoted a few General Managers from research and engineering leadership roles because we knew their talent and capability; but when we went outside to hire a GM because nobody inside was ready, we hired someone who was already doing a similar GM job in a similar company. In other words, if you are an unknown, it is very hard to change companies, industries, and positions all at once. Chances are good that someone else already has the exact qualifications for the job you want, and you will be left behind.

Do Stuff That Matters to Your Company

In general, it is a good idea to *stay close to your organization's mission because impact and visibility are keys to advancement.* I may have done a terrific job on coal research for my first job at GE, but nobody really cared because it had no impact on the company bottom line.

If you are in a corporation, growing sales, cutting costs, driving a key merger, and adding to the top and bottom line are really what matters. How about university service? Most likely, what matters is bringing in research dollars and enhancing the university's reputation through strong peer-reviewed publications of cutting edge enquiry, along with winning national and international awards. Being a good teacher is OK, but may not be the key driver in a research university. But for a small college, you bet teaching matters.

Do you work in the government? Do something that helps get the big guys re-elected and it won't hurt your career a bit. For non-profits— most likely raising stature, and the money that comes along with it, are still the key.

And no matter what type of organization you work in, bringing home bad press is never a good idea, except perhaps in show business where bad publicity is better than no publicity.

Who Are You Kidding?

Sometimes folks who asked for my advice were way off on their plans and goals as compared to reality and I tried to gently, but candidly, ground them.

One colleague of mine, for example, was working as VP of research at another company and spoke with me about his career plans. He felt that his next step was to lead a business, a "P&L" he called it which stands for profit-and-loss. It is a basic business unit entity, usually led by a general manager, and is an important step toward becoming CEO. I had to tell him that, however competent he might be in his current role, and even how capable I thought he would be in the P&L job, I felt it was extremely unlikely that he would be given such an opportunity coming directly from R&D with no commercial (sales or marketing) experience. I don't like to rain on people's parades, but realistically, if that was the role he desired, I felt he needed a step in between and he should spend a couple years in sales or marketing to be a truly viable candidate.

I once had an employee in his mid-fifties who told me he wanted to be CEO. I think he wanted to be CEO because of the ego, maybe the money. But he didn't think about the nature of the work or whether he would be any good at the job. And aside from that, since he was a research scientist and had been a research scientist all his life, I had to tell him there was only one way he would ever become a CEO, and that was to start his own company. Quite frankly, even if he was capable of being a great CEO, in all likelihood, by the time most anyone is fifty years old with no business management experience, nobody is going to hire him as a CEO, and it was probably too late for my pal to get the necessary experience, at least with our company at the time.

On the other hand, a laboratory technician in his twenties, with only an associate's degree, came to talk with me about his career, and told me he also wanted to someday become a CEO. I guess being a CEO must be considered nice work if you can get it. And actually, this was a kid with hustle and a great attitude. If I squinted, I could picture him someday as CEO. As we talked about how to realistically get there, I felt he absolutely needed to get a BS degree in engineering, and that he could move through Operations further and faster than R&D. In R&D, it is very difficult to advance without making a salient technical contribution as well as earning the respect and credibility of the scientists and engineers. Almost certainly, the R&D leader would need a PhD in one of the sciences, or maybe a master's degree in Engineering and Business. But in Operations, on the other hand, he could quickly become a shift leader, and could work his way to becoming Plant Manager. From there, if he was really a great leader, I could see a progression to COO and finally CEO. It was at least possible. He might need to move from company to company, region to region, industry to industry, but at least I saw a way. I also suggested the entrepreneur route for him to skip all those intermediate steps. He did, by the way, complete his degree and his future is brighter than ever. What is the likelihood of his becoming a CEO at a significant corporation? Probably less than one percent. But it isn't zero, and he has a path. Good luck, my friend.

Some of my R&D employees have told me that they wanted to become General Manager or Global Business Manager or Operations Manager. Again, they may be perfectly capable of doing these jobs well, but they will not easily move from being a researcher or engineer into these positions, and it is even a stretch to go directly to one of these spots in one step from being an R&D Manager. Yes it happens, but do you want to bet your career on long-shots? I don't think so.

Although I am especially familiar with R&D individuals and careers, similar big steps are unrealistic in all disciplines. Sure it happens, but don't bet your future on it. Take big steps and little steps in the direction you want, and eventually you will get to where you want to go.

Paying Your Dues

Let me remind you about the brief discussion we had in Chapter One about paying your dues *versus* taking a job you don't want in order to qualify for a job you do want. Everyone does have to pay their dues, but that usually involves a *bona fide* apprenticeship, where you learn the things you need to know in order to succeed at the next level. If you really want to be a General Manager, but you loathe being a Product Manager who reports to a General Manager, you really need to question yourself why you feel that way. Perhaps you consider the level of that position beneath you—maybe lower than your current position. Maybe you consider it a step backward. I would say "nuts" to that. Levels are artificial boxes on a page. Most compensation systems have enough spread such that taking a step down or over will not hurt your pay a bit. If your ego is bruised by moving from a level-twenty to a level-nineteen job, I think you are focused on the wrong thing. Focus on what you can learn, how you can grow, what impact you can make, and the dollars and titles will follow, as we will see in the next chapter.

Look around the brass in your organization and I think you will find that most of them have rotated through several assignments and organizations to get where they are. By the way, it is a great way to attract visibility and to find good mentors to simply ask the most successful guys in your organization something like, "Hey if you have a few minutes, would you be willing to share with me how your career progressed so well?" They might blow you off, or think that you are being a sycophant, but if you are sincere, you might learn something very useful and they might be happy to share with you their hard-knocks stories.

One consideration for advancing your career includes a willingness to relocate. If you are unwilling to relocate, you are often limiting your career opportunities. Maybe your company has a big job in Asia that matches your career plan perfectly, but your American family has other ideas. You could be stuck and your career derailed. I recall being in this situation myself following my divorce. Staying near my kids took precedence over my career. Sorry folks but you make your choices in this world and some are not compatible with your career aspirations. These days, many companies are very accommodating to work-life balance, but please be realistic about what any company can do for you. Commuting to Asia or relocating temporarily is expensive and limited to special situations. Most organizations prefer to hire locally where possible. Often the folks at the top of major corporations are blessed with a spouse who willingly packs up and moves the family across the globe, often at two-year intervals and with little advanced warning.

Many people change companies or industries because opportunities in their current company are limited. R&D people leave to become professors or fellows or patent agents or to change industries, to get promotions or for personal reasons such as to follow their spouses or take care of an aging parent. All this is understandable. I personally changed companies twice—once my choice and once theirs. Relocating is a fact of life. Restricting yourself to one location is a serious career-limiting situation.

As I mentioned in Chapter One, I recommend against taking a job that you are sure you won't like in order to eventually get the dream job. Life is too short and uncertain to be taking roles you know you won't enjoy, and there is no guarantee you will move on successfully. In my opinion, it will be harder for you to succeed in a job you don't like, and it is certainly an important criterion for promotion that you are successful in your current assignment.

On the other hand, how do you know you won't like something different? People might think to themselves that they would hate to live in Taiwan or Malaysia or Ireland, would hate to work in Quality or Sales, could never work for a woman boss, would never want to have unionized employees, and so on. Are you sure? Be careful to not dismiss opportunities, including lateral moves that offer personal growth. If your supervisor makes the recommendation for you to take a lateral assignment, she probably sees something in you and your career that you don't. Often I hear people dismiss opportunities without really thinking things through. Maybe you will find out that you like your new job, after all. I find that there are interesting tasks and assignments in all walks of life, and that good people tend to gravitate to assignments that offer the biggest growth and impact, regardless of the specifics.

If you turn down more than one or two opportunities, you may get labeled as a "tree-hugger," someone who is unwilling to take a chance on change. Pretty soon they will stop asking.

Staying In a Job You Love Forever

There are numerous jobs that are so intrinsically rewarding, that many individuals working in these roles don't really care for advancement, other than furthering their mastery. I would include teaching, all of the arts, certainly research and development and engineering, the judiciary, police or private detective service, farming and gardening, cooking, coaching, athletics, entrepreneurship . . . I'm sure you can think of dozens of jobs where people are just happy to do what they do and aren't looking for titles or recognition. I love people like this who have found their niche and whose career means nothing more than expanding their capability within the role.

One of the things that I liked so much about GE's central research was that there were no levels for the researchers, only the managers. You were considered at the top from the very start. At many companies, R&D organizations—like every other department—come with a plethora of levels such as Engineer I, II, IIII, Associate Scientist, Scientist, Senior Scientist, Fellow and so on. We'll hear more about Titles and Levels in the next chapter.

As you would expect, in organizations that emphasize levels and hierarchies, people naturally become more focused on the titles and levels and compensation, and the inevitable inequities associated with their disbursement, than on the intrinsic work. I feel this is often to their (and their organization's) detriment. I am sure that well-meaning managers and human resource departments feel that levels and promotions are motivating and inspiring for individuals. On the other hand, Daniel Pink describes in *Drive* how these distractions turn off creative, knowledge workers who often perform their best when unfettered with extrinsic reward structures, provided they have sufficient income and status, along with freedom and autonomy, to begin with.

So to summarize, take a deep look at what you really want and are good at, and where you want to go. Determine if it is realistic, and if so, think about the possible steps along the way. Get advice and counsel from people who are already in those jobs and from people you know and trust. Be prepared to roll with the punches as the unpredictable will surely happen to you; and by all means, get started soon.

Chapter Five

Titles, and Levels

BACK IN 1995, WHEN I WAS working at GE's Corporate Research and Development as a Program Manager (PM), the head of R&D managed to get all of the PM's promoted to GE's Executive Band (EB). There were about a thousand or so EB's in the company.

Being EB entitled you to a few things, though not really all that much. You became eligible for an annual bonus that started at a few thousand dollars per year and which went up each year depending on your performance and your division's performance. The downside was that more of your compensation was "at risk" and you got smaller base salary increases. Also you became eligible for a special pension supplement if you stuck around to age sixty, though in actuality, very few people in GE stick around to sixty. Also every few years, EB's got a special option to bank up to half of their salary to defer taxes at a very good fixed interest rate on the deferral.

The upside of joining the executive band was greater income of a few thousand dollars a year or more as time went on. On the downside was that you were now thrown into a pool of all executives for performance evaluation purposes, and at that time GE was committed to dropping the bottom ten per cent of performers at each level in order to add fresh new blood. So you could easily move from being the top performer at your previous level to becoming the bottom performer at your next level. It was generally better to get a higher rating at a lower level than a lower rating at a higher level in terms of both salary increases and job security. Surprisingly, the higher the level, the greater percentage of people who got let go. At the officer level, the turnover was a whopping thirty per cent each year, meaning that the half-life of a VP was less than three years!

In September of 1995, just a few months after the promotion to executive band (EB), I got the opportunity to move from central research to our Silicones business. It looked like this was a good career opportunity in terms of growth and experience, but with a hitch. Each business was allotted only a small number of EB positions, and it practically took an act of God to increase the number. My new boss-to-be told me he could not promise me an EB position, and asked me if I would take the job, anyway. There would be no difference in the compensation either way. I told him that, if he promised me that he would do everything he possibly could to get me the EB, then I would promise to take the job—with or without it. I think this may have been a litmus test to see how serious I was about the opportunity. I was indeed quite serious. My answer was a good one because he was able to get me the EB slot, after all. I got the position, the title, and the goodwill that went along with it. It turned out to be a terrific move for me. I enjoyed some success there, and gained valuable experience which helped me move on and up in the years to come.

The key is, I was willing to take a job that could have been at a lower level because I valued the personal growth and impact more than a title.

Now let me contrast that with another coworker at central research—someone I'll call Karl since this is a true story and people will probably try to figure out who I am talking about. Karl was in the same boat as me—a newly appointed Executive, and was offered a position in another GE business with a situation similar to mine; the proposed job was great but they could not promise Karl the executive band. In contrast to me,

Karl made it clear that he would **not** give up his EB for the sake of the job, even though they assured him that they would more than make up the difference in cash. Unfortunately, the other business was unable to create a new EB slot. As a result, Karl turned the offer down, even though it was a fantastic career opportunity. Naturally, the head of that business raised holy hell to the R&D VP, claiming that level inflation cost him and his business a good employee. As you can imagine, Karl was not terribly popular with his VP after that. Guess what—ten years later Karl was still in the same job at central research that he was in back then. I was on my second officer job making probably four times what he was.

Opportunities don't necessarily keep coming forever, and deciding which ones to jump for is tough. My advice is to focus on the job content, your personal growth, and the visibility and impact you might have, and forget about titles, grade levels, and to a certain extent, compensation. I probably would not go down in compensation unless it was just the dream job of a lifetime, or if I was very sure that the growth or impact would be forthcoming. And realistically, most hiring managers would not expect you to take a position that paid less, unless you were moving to a much lower cost region, or because you had to take a job due to unemployment or a spouse's relocation, for example.

Leading High Performers or Misfits

There is one other interesting anecdote about my move from central research to Silicones in 1995. I had originally met with Silicones' process engineering director to explore career opportunities there. He told me that there was not a good fit for me, but that there were more changes planned in the next few months, so "hang loose." I did.

Indeed a couple months later, I got the call that there was an opening to lead the Silicone Sealants group—the team that develops tub and tile and construction caulk known as RTV or "Room-Temperature-Vulcanization." I spent all day interviewing for the job, meeting with the RTV R&D team, the RTV manufacturing team, 'and the RTV business manager.

At the end of a great day, the head of R&D brought me in to wrap up. My future boss said "I have only one question for you—If you had the choice between leading a high-performing team with great talent and energy and enthusiasm and fantastic projects—OR—leading a completely dysfunctional team of misfits who have nothing but complaints and problems—which would you choose?" In a millisecond, I enthusiastically said, "Give me the misfits, any day!" He smiled broadly and said "OK then you have the Rubber job, not the RTV job."

That was another good answer. When you take on a new job leading a high-performing team, perhaps the job seems easier at first, but it is always tough to follow a good leader and make things better. Following a superstar, you tend to not get credit for your accomplishments, and it is harder to make a big difference. You are far better off taking on a disaster assignment, because people will expect less and tend to blame the previous boss for a while which will give you air cover; and moreover, if you are successful, the contrast will be stark. Plus it is more fun to turn things around rather than just stay the course.

The funny thing about the "misfits" I inherited was that the boss either had a bad misimpression, he was exaggerating, or maybe he was just testing my mettle. The rubber group was terrific and especially had an incredible group of talented technicians—by far the best group of technicians I have encountered anywhere before or since. All I really needed to do was make sure that they got good assignments and were left alone. I got rid of a couple of "bad apples," cut the technicians loose, and in a short while the Rubber team became the best department, and I got lots of undeserved credit.

In a similar way, consider the successful San Francisco 49ers of the 1980's and 1990's. In 1979, Bill Walsh took over as head coach of the 49ers who were 2-14 the previous season. By 1981, they won their first of five Super bowls. What an amazing turnaround. George Siefert eventually took over from Walsh in 1989, and though he won two Super bowls in the next few years, he never really got the respect that he might have had he turned around an unsuccessful franchise. Bill Parsall's on the other hand took the lowly Jets to the Super bowl a few years after taking over as coach, and, as a result, he got an express ticket to the Hall of Fame.

So my advice is, don't worry too much about job titles; take on assignments that offer you the most growth, fun, and the opportunity to make the biggest impact—and that will accelerate your career and job satisfaction way more than a fancy title and big office.

Chapter Six

Choosing the Right Organization: Big Fish/Small Pond or Small Fish/ Big Pond

To a certain extent, your skills and interests, training and education will select your organization for you. If you are a social worker, probably you will work in a government agency or maybe a private clinic. If you are a nurse, you will likely work in a hospital, clinic, or in private homes. Police or fire service? I think you will be a public employee. If you chose teaching, you will likely work at a school . . . call me crazy. I am often amazed about teachers who complain about their jobs, when they probably knew better than most of us what they were getting into before they started their careers. But let's not go there, now. And of course, teachers can choose among urban, suburban, or rural school districts and schools of various sizes. Professors can choose among major private research universities, large state colleges, small private colleges and so on.

In our case, we chose science, technology, engineering, innovation. Still, we have lots of choices on where to work.

More and more these days, we can actually choose what country we work in, even if we are on a foreign or temporary visa. Yes Tom Friedman is right—*The World Is Flat* a terrific book to better understand the impact of globalization on our work and lives.

For many of us, we have a choice between the public or private sector, profit and non-profit. We can choose among public and private corporations, large cap, mid cap, small cap. We can select among local, regional, and global organizations. We can choose focused companies or broad conglomerates. How do you begin? Do you just roll the dice for your first job and see what catches and go from there? Most people do it this way, actually, because they don't have a lot of choice. Jobs are not exactly growing on trees these days and most of us take what we can get and then go from there.

Some organizations consist of just one person—you! According to the US census bureau, three fourths of all US firms have no payroll and consist of "self-employed persons operating unincorporated businesses."

Take a look at the table below from 2008. I'm sure it is somewhat different today, but similar in many ways to the world of today and tomorrow.

Business Size	Number of Employees
Firms with 1 to 4 employees	6,086,291
Firms with 5 to 9 employees	6,878,051
Firms with 10 to 19 employees	8,497,391
Firms with 20 to 99 employees	20,684,691
Firms with 100 to 499 employees	17,547,567
Firms with 500 employees or more	61,209,560
Firms with 500 to 749 employees	3,681,760
Firms with 750 to 999 employees	2,617,087
Firms with 1,000 to 1,499 employees	3,720,654
Firms with 1,500 to 1,999 employees	2,653,392
Firms with 2,000 to 2,499 employees	2,011,244
Firms with 2,500 to 4,999 employees	6,726,611
Firms with 5,000+ employees	39,798,812

It's an interesting distribution. There are a lot of folks working for big employers with 5000 or more employees, and there is another peak around companies with 100-500 employees. But there are still big numbers for firms of all sizes, and behind each one of those numbers is a real person just like you who chose to work at that job and that organization for a myriad of reasons.

For a small company with just a handful of employees, there is the joy of *getting* to do everything—making the sale, paying the bills, doing the design, assembly, manufacturing work, handling the paperwork, setting the strategy, creating the IT, reaping the rewards. But there is also the aggravation of *having* to do everything—closing the sale, paying the bills, reaping rewards that don't pay what you might want. It is a high risk, high autonomy, high reward situation. Everybody knows everybody and relies on each other. It can be a family, and not all families get along sometimes. And remember that nobody becomes an entrepreneur to avoid work; you will need to bust your hump.

Bigger organizations allow some specialization. Maybe there are a few rainmakers who do all the bidding and selling, while everyone else is involved in some form of production or fulfillment. Maybe one person does the books, another the logistics, two folks do maintenance and so on. Still close knit with a lot of camaraderie, smaller organizations may not have the capital to do everything they want, and may frequently be out on a financial limb.

As you get up to the five-hundred or thousand person organizations, there are bona fide departments and specialization. Still small enough to know almost everybody and where each individual can make a big difference to the company's bottom line, now the organization is big enough where there are truly specialized functions such as engineering, manufacturing, finance, human resources. You can focus more on the areas of your expertise while allowing others to do their part.

Still larger organizations now begin to depend much more heavily on extensive management and systems to coordinate major activities that complement each other. Engineering will need to flange up with sales and marketing. Manufacturing will need to get orders from Sales, and hand off finished goods to Logistics. Quality systems and Supplier sourcing will play key roles. There might be a large Finance department and even Legal. Companies of this size are often publicly traded, and will have investor relations departments and government compliance teams, internal audit, an outside Board of Directors. Individuals in this size company can get lost in the organization as it is harder to stand out and make a large impact. But due to the highly managed set of specialized functions, individuals are also less exposed to the fluctuations of business cycles and may have more job security than found in smaller companies. And certainly the larger companies can afford state-of-the art resources and facilities that can streamline your work. With larger organizations comes better services—a top notch IT department, a library, great internal communications, a cafeteria, maybe a gym and a daycare on site. Comprehensive benefits come with the package. And along with all these great systems comes bureaucracy, petty internal politics, and a pace that may be too slow for you.

I have not worked for the public sector, other than as a summer employee at the Argonne National Lab, and as a volunteer firefighter and Fire Commissioner. Nor have I worked for a non-profit other than serving on the Board of the local United Way where all of our agencies were non-profit. But people being people, and organizations being organizations, I would suspect that, just like the private sector, the bigger and more complex the organization, the more political, bureaucratic, and slow they become, though with the advantages of better pay, more specialized departments, improved facilities, and greater job security.

Big Fish Small Pond?

Regardless of the size and shape and industry you join, are you the kind of person who would you rather be the best player on an average team or an average player on the best team? Would you rather be a big fish in a small pond or a small fish in a big pond?

It is important, no matter what size of pond you swim in, to stand out. Often corporate human resource departments, and their compensation systems, establish distribution guidelines (see Chapter Eight on Performance Reviews) for performance ratings that tend to limit the number of top performers who get the highest raises and bonuses. If you stand out, you can count on better compensation and more career opportunities.

Chances are that the bigger the organization you join, the harder it will be for you to stand out. If it were just a game of numbers, it is likely that the bigger companies simply have more folks for you to compete with. But big companies tend to also attract the best and brightest candidates to begin with, recruiting at the top schools and offering great compensation and work conditions. They have a virtually unlimited pool of applicants, making it easier to cull some of the weak performers, knowing that they can easily be replaced.

Still, if you join a great and reputable institution, and you manage to stand out, you are truly special and have unlimited potential anywhere.

If you are part of a strong team, it may also be difficult to stand out. On the other hand, more rewards, recognition, and visibility will come to members of a successful team. And being on a winning team and associating with winners and stars has its intrinsic rewards. Your coworkers will make you better. And no matter how good you are, if your team is not succeeding, it will be hard for you to feel very good about yourself or to get high ratings.

My attitude is that, it is generally better to be among the winners and take your lumps for being average, but by no means is this the best strategy for everyone.

When I first went to Stanford from little Willoughby, Ohio, I *knew* I didn't belong there. All these kids had 4.0 grade point averages, were valedictorians, had 800's on their SAT's and were Olympic athletes or published scientists as well. And I was right! Boy I struggled to keep up. It took me all of four years but when all was said and done, I finally felt like I belonged there.

My point is, if you surround yourself by people who are better than you are, what the sociologists call your reference group, yes you may not exactly fit in, but you tend to reach further and try harder to catch up or keep up, and in the end, it makes you better for it. We all want to belong. But it isn't easy on the psyche; that's for sure.

And the bad news is that, when you take the hard classes with the strong students at the top colleges, you can get a lot of C's. When you play for the Yankees and hit 0.280, you will probably warm the bench or bat seventh and nobody wants your autograph. Are you self-confident enough to deal with that? Are you able to keep things in perspective? Some people aren't.

In my own case, I remember when I was just seven years old, I was a pretty good ball player for my age. In tryouts for the Little League, I was one of just a few seven yr. olds that started out in the minor leagues; most sevens were in the pee-wee t-ball league, while eights, nines, and tens made the minors, and ten-to-twelves got into the majors. But the sad case was that, because I was out of my league, I got very little playing time or coaching that first year in the minors. I mostly rode the bench, pinch hit, and came in for late innings where we had a big lead. In this case, I probably would have been better served had I stayed in peewee and been one of the stars and gotten a lot of playing time.

So for many people, it is actually better for them to be the best in a smaller universe. It is just too hard for them to cope with slipping into the pack.

In my high school, there was a truly exceptional, brilliant young man. He won every academic award we offered—science, English, mathematics, several foreign languages. He went on to Caltech which at the time was the most selective and challenging college in the world. I think only about two hundred kids were admitted each year. Many years later, I would become a graduate student there, where it was nowhere near as competitive as it was for undergraduates. I heard that my high school friend had struggled there, as did almost all the undergrads. It is truly a brutal academic environment. I am sure he eventually went on to a great career somewhere, though I wonder if he had gone to a more normal University, he surely would have gotten a 4.0 grade point, won all sorts of awards and fellowships, and would have moved onto the next level brimming with confidence and energy. My point is that sometimes, life in the fast lane comes with a price that is too dear.

Chapter Seven

Doing the Job Well

ASSUMING THAT YOU ALREADY HAVE A job or are about to embark on a new job, eventually you will be expected to perform well, to execute on the tasks and meet expectations. The reward if you succeed usually means that you will be faced with tougher challenges and higher expectations. A series of progressively responsible positions is really the essence of what makes up a career.

One way to keep your career on track is to continue to succeed, to execute, and to deliver the goods. As this becomes harder and harder, the likelihood that you will miss from time to time increases. This is not necessarily the disaster it may seem at first glance, and coping with setbacks is a skill you will need to master. We'll hear more about this in Chapter Twelve.

There are any number of reasons why certain ventures don't always work, and nobody is expecting perfection, particularly as the risks escalate. Presumably, as long as you busted your hump, made reasonable choices, and failed for logical reasons that may have been out of your control, you will probably be given a second chance, maybe even a third. Someone who consistently swings and misses will probably get off the track and will need to restart his career somewhere else. This is also not necessarily a disaster presuming you learned the reasons for failure and won't repeat those mistakes.

Given the alternatives, succeeding is generally better than failing, though I like to hire people who have occasionally failed and who have learned from it. It is relatively easy to succeed all the time if you take no risks. I can almost always assure you that I won't lose money simply by keeping it in the money market account. But slowly I will have my purchasing power eroded by inflation and devaluation and taxation. Similarly a career with no risks, like any business, is a certain loser. Learn how to take good risks and mostly deliver the goods and you will be OK.

If you are in an organization that punishes risk-takers who fail, you might consider an exit strategy. Quickly, people in organizations like this get the message to lay low; a decline is inevitable as more aggressive competitors begin to take your share of market. On the other hand, if you see risk takers get rewarded for missing, you have a special leadership team and a *bona fide* future of growth, both personal and corporate.

I will never forget sticking my neck way out at GE Lighting, trying to develop a clear rubber-coated halogen light bulb that had outstanding cost and performance, but with a one-in-a-million change of exploding. The rubber coating was developed to contain the glass in those rare instances. After a couple years of taking on—and succeeding—in overcoming design, assembly, performance, and cost obstacles the business finally killed the project in response to an economic downturn. Despite our disappointment, the company rewarded many of the key participants with stock and cash. Similarly, after working so hard to acquire and integrate Honeywell to the GE family, after the deal fell through for political reasons, Jack Welch sent many of the due-diligence and integration leaders cash, stock, and a nice personal note of appreciation. And nobody felt that his GE life was totally derailed because the deal fell through.

What Kind of Person Are You?

You know the old joke—"there are three kinds of people in this world—those who can count, and those who can't." OK there are really two kinds of people that I want to talk about now.

The first kind of person tries to figure out the bare minimum effort they can get away with—the least effort needed to pass the test, or to not get fired, for example. Sometimes these kinds of people go to greater efforts to get out of work than is needed to do the actual work itself. They secretly fear or resent the fact that they might be doing more work than the next person for the same—or even less pay. I suspect that some economic and compensation systems drive toward this behavior—i.e. ones based on seniority or need rather than output or performance. And well-meaning parents may even encourage this behavior pattern by paying for good grades, for example. This person might be thinking "I work twice as hard as so-and-so and she gets the same pay . . . Am I stupid or what?" This person is extrinsically motivated.

You never want to be this person. You will never utter these words: "That's not in my job description."

The second type of person, in contrast, tries to figure out how to get maximum output and results, regardless of the output or compensation of others in similar roles. This person tries to find out how to do *more*, accomplish more, not less, in a given day or week. This type of person is intrinsically motivated and gets real satisfaction out of hard work and the resulting output. This person usually works harder than the extrinsically motivated person despite the inequity of the situation.

If you want a great career, make this person you.

A great book on motivation that I mentioned earlier is *Drive*, by Daniel Pink. Pink describes rewards and compensation systems that completely undermine the intrinsic human enthusiasm and curiosity and passion for meaningful work. I highly recommend that all supervisors and parents read this book.

What I find interesting is that, despite any imbalance that the second "intrinsic" person may feel about rewards and compensation, he is usually happier and more energetic than the first type of "extrinsic" person, taking pride in a job well-done and the satisfaction of going home beat after putting in a hundred per cent. Eventually this person does get his external reward and recognition—promotions, raises, and so on. People really notice.

Doing Your Main Job Well

Take a moment to think about your current job, a past job, or even a future job. What are the key things you need to get done? What are the goods you need to deliver? What commitments did you make? What are the expectations for you?

If you don't really understand what is expected of you and what are the key measures, then start here. You need to have frequent discussions with your supervisor to make sure you truly understand what she is counting on you to deliver, and also to touch base frequently with her to make sure you are on track. Sure it may be officially your boss' job to initiate this, but remember my admonition above about "That's not my job." Make it your job. You simply must always be working on the right problems. Often, this is a moving target, and somebody may have forgotten to send you the memo.

You not only need to understand what is expected of you, but also why. We call this "line of sight" and what it means is that you have a direct line of sight between what you are expected to do, and how it will impact your organization assuming you are successful. It is amazing how many people get this part wrong. Don't be one of them.

For example, let's say you are a quality engineer, and your job is to address product performance excursions. It seems obvious to me that poor quality leads to lost customers, and this affects your company's sales and profits. Your job is to make sure that doesn't happen. The biggest thing you can do as a quality engineer is to make sure that you nip quality issues in the bud permanently. If a quality excursion does happen to slip through, you then must make sure that you fix the problem and ensure that the customer is apprised and satisfied with the fix and is remunerated for his troubles. Can you see the line of sight between your work and the impact on the company? I hope so.

Continuing with our example of a quality engineer, even though it is clear that your role is to prevent and fix quality issues, less clear are your specific measures. The boss may measure you on response—time to the customer, or the number of days between the opening, and closing of a customer complaint. You may not have a specific metric for delving deeply into the details of an excursion and coming up with a permanent solution, but I have a feeling that if you expand your role to include this, and you deliver, that you will get a lot of positive attention. Do you see how delivering the goods and expanding your responsibilities go hand in hand? It is important to focus on the formal measures for your job, but it is even better to focus on the meta-measures—things that will really make a difference to your organization.

Starting a New Assignment

Assuming you have taken the initiative to make sure you understand your new deliverables, timing, and measures, what are some strategies to meet and exceed expectations?

One really good approach comes from the book, *The First 90 Days*, by Michael Watkins, which I strongly encourage you to read and reread with each new assignment. The book recommends that you spend the first ninety days on a new job getting the lay of the land, learning the expectations, and negotiating hard for the resources you need to become successful. It is painful to be given an assignment and then find that you lack the time, the materiel, the staff, and the resources to succeed. The more you can negotiate this in advance or early on, the better. The book gives you strategies on how to broach the subject effectively.

Meeting Expectations and Beyond

I would not recommend "sandbagging" which is to exaggerate the degree of difficulty of the job, then under-promise and over-deliver. I know this is a common strategy to cover your behind (CYA) in case things don't work out. You will quickly get a reputation as a sandbagger and people will not take your estimates seriously. If you exaggerate the challenge too much, the company may decide to stop doing your project, altogether. It is better to be realistic and speak in terms of probabilities and risk. Allow yourself to take on higher risk projects if the impact is great enough, and negotiate adequate resources and create contingency plans, where possible.

Assuming you have the expectations, measures, timing, and resources, then you then need to work diligently and effectively toward meeting the goals and objectives. Nobody will really care about the other things going on in your life at, and away, from work. There will be numerous distractions—required human resources and safety courses on sexual harassment or blood-borne pathogens; maybe a temporary crisis will arise that will require you to drop everything to assist. There might be a key resource who takes ill or quits or gets pulled away. A supplier might miss a delivery. A customer might temporarily move your project to a lower priority as they face a crisis of their own. A black-swan event (see Nassim Taleb's great book *The Black Swan* about the impact and role of low probability, high impact events) like a tsunami or war might derail your timetable.

It is easy to get angry and frustrated when your resources get taken from you. Keep your cool and cheerful outlook. It is a way of life in this world, and how you cope with disruption is something people will be observing in you.

It is far better that you build in contingencies into your plans, and find ways to keep always the ball moving in the right direction. If need be, you and your team will work have to work longer hours, weekends, skip holidays, take on expanded roles and do whatever it takes to get the job done on time. Do this during a crunch and you will be a hero. Work smart; work hard. Find a way. It is far better than having to make explanations that always sound like excuses. Despite all the obstacles and unanticipated problems, when all was said and done, you found a way to deliver. With a smile. Can you have this attitude? If you don't, someone else will, and they will get the promotion. Sorry—that's life.

If you are chronically overloaded, working way too much and sacrificing your personal life, you may be over your head, in the wrong job or organization. Or it may be that you are working inefficiently on things that don't matter, and not focusing on the key elements that really benefit your company.

The Pareto Principle

I believe that some people are significantly more efficient than others. I suspect that they are intimately aware of the Pareto Principle. Do you know about the Pareto Principle? Wikipedia tells us that "it was named after the Italian engineer, economist and mathematician Vilfredo Pareto, 1848-1923. The Pareto principle (also known as the eighty-twenty rule, Haddad's Theorem, the law of the vital few and the principle of sparsity) states that, for many events, eighty per cent of the effects come from twenty per cent of the causes. For example, Pareto observed that eighty per cent of income in Italy went to twenty of the population. It is a common rule of thumb in business; e.g., eighty per cent of your sales come from twenty per cent of your clients." And eighty per cent of your sales come from twenty per cent of your products. And eighty per cent of your successes come from twenty per cent of what you do. I'm sure you get the gist.

So of course, my Dilbert mentality says "Why not just skip the eighty per cent that is a waste of time?" Is the secret of life that simple? Well— yes in a way. The most successful people I know have a way of focusing on the important things and letting the other stuff slide or finding ways to delegate it to others. I know of many very busy and hard-working people, but lots of them don't tend to get very much done—at least the stuff that matters. I get real scared when I see people with appointment books that are overfilled, showing every minute booked for the next couple weeks. Maybe they just don't know how to tell the vital twenty per cent from the useless eighty per cent. How do you do this?

Let's begin with the end in mind, as Stephen Covey tells us in his business classic, *Seven Habits Of Highly Effective People*. What are we really trying to accomplish? Go back to those line-of-sight objectives we talked about in the paragraphs above, and laser—focus on these. Sure there are distractions—keep them to a minimum. Yes others book up your calendar and you have trouble with interruptions, meetings, and multitasking. Who doesn't?

One of the most effective strategies I discovered was coming into the office way early. You always get a great parking spot, and everyone notices your car in the first spot and your light on and your half-full coffee pot when they come into the office. People from Asia or Europe would often call me to leave messages and were shocked to find me at my desk at five or six AM. It sends a good message and creates good image for you. I know this might not be possible, but it is amazing that you can almost get a full day's work in between six and eight AM when you are free from interruptions. Similarly, staying late or working a Saturday morning every now and then can give you a major boost in catching up on things that may have slipped. I prefer coming to the office rather than working from home; it just gets me in a work mindset and I am free from the distractions of home. Doing some reading at home is a great idea, especially in areas that enrich your knowledge and expand your capability.

Where possible, skip the stuff that doesn't add value, and aggressively pare these things from your program. You might get in trouble from time to time if you blow off some of the administrivia that pervades all organizations, but if you continue to deliver big, all sins will be forgiven. And if you miss often, keeping your nose clean won't help you a lot. You do the math. Now spend your efforts where they matter.

By the way, a terrific book to study is, *Execution*, by Larry Bossidy and Ram Charan, who take a 'no-excuses' attitude for getting the right stuff done, and how.

As a supervisor, one of the greatest things you can do is clear the deck for your employees so they can focus on the important tasks instead of getting bogged down in meetings.

Lots of Other Ways to Contribute and Have Impact

I really enjoyed the book, *You Can Observe a Lot Just By Watching*, by Yogi Berra. Yogi was a phenomenal baseball catcher and manager. In seventeen seasons with the Yankees, he played in fourteen World Series and won ten of them. He knew that the World Series was the Yankees' line of sight goal every year. The book is mostly about teamwork.

One of the things he talked about is how he and his teammates always looked for ways to help their team win. He talked about going zero-for-four, that is, not getting a single hit in four at-bats in a game, but still finding a myriad of ways to contribute. Obviously, he could catch a great game, call all the right pitches, move the players into proper position, back up throws to first or third, throw out a player stealing or bunting and so on. I think this is a good philosophy for all of us.

We talked about what is your main job—delivering on those line-of-site, high impact items that you, your boss, and your company is counting on. Never lose sight of what is it that your teammates are counting on you to do. Develop a new product? Help your sales counterpart to close a sale? Solve a critical quality issue? Most of us have a pretty good idea what our main job is and how it directly, or indirectly, affects the organization's success. But now let's think about what your other jobs might be.

I bet you could identify a dozen ways to contribute besides your main job. Almost all of us have specialized knowledge. Sharing that knowledge is one of the most important things we can do. The more each of us knows, the better we all are. Have you sat in on brainstorming sessions? Be thoughtful about your issues and suggest solutions. Nurture a new junior employee.

I bet you have looked for ways to save money in your areas of purview . . . but what about other areas? Do you know a cheaper hotel in Singapore? Have you heard of alternative source of a key raw material for your company? What little tidbit of competitive knowledge did you hear through your grapevine? Do you even have a grapevine? Can you use it to help find a great new employees?

Have you read a good book or article that you think others would value? Did you find a cute trick in some accounting software? Did you clean up your mess and someone else's even if it was really their job to do so, but you just happened to have a few minutes in between tasks?

Have you spent a few minutes with someone in other departments because you identified an issue or opportunity they might be interested in? Did you carpool to the airport to save parking costs? Did you cheer up a dispirited coworker? Did you stay home when you had a cold because you didn't want to infect the whole team? Did you learn how to add 3D animation to a Power Point presentation? Did you see a patent that you think infringes your company's prior art?

If your company has a customer's service hotline, even if you are unqualified to answer customer complaints and issues, spending a few hours just listening in on the hot line is a fantastic way to learn what is truly important to your customers—in other words, a direct line of sight to issues that matter to your organization's ultimate success or failure. Try it for an educational and humbling experience.

I think you get the point. You may strike out from time to time when developing that new product because the competition got there first. But you can still find a hundred ways to add value, if only you look for it and go for it.

Are you confused? The Pareto Principle says to focus on the twenty per cent of things that really matter. Now I am telling you to look for other ways to contribute. The key is to do both. Make sure you get your main job done, but also look for low-hanging fruit—quick and easy opportunities to help your organization in other ways. If you do both, you will have a great future anywhere. And when you do find small ways to contribute, make sure you record them in your files—you and everyone else will forget them unless you keep track, yourself. As we will see in the next chapter on Performance Reviews, small edges make big differences in an environment where the boss has too many people and can't track everything they do; where little snippets of exposure shape opinions and outcomes.

Chapter Eight

Performance Reviews

MOST ORGANIZATIONS HAVE SOME SORT OF formal performance review process, usually done on an annual basis but occasionally more frequently. Performance reviews affect compensation and promotion, and provide a window into how you are regarded and what will be your future trajectory. Sadly, performance reviews often lack candor or fail to offer meaningful advice. Despite this, we will look for ways to read between the lines.

To start, let's look at the basic the nuts and bolts and maybe even reveal some of the warts and dirty little secrets of the process.

By way of preamble, most people—employees, managers, human resource professionals—believe that a performance appraisal system is a good and necessary thing for companies to succeed, thrive, grow, and to fairly reward and recognize the people who made it all happen. I would bet that most of you would prefer to be employed in a system where people who worked harder and delivered more impact got paid more. And probably you feel that no system is perfect—that appraisals are often arbitrary. There may be cronyism and bias, and certainly some people benefit from cake assignments and easier bosses. Many may get carried along by others who do the heavy lifting, while other bright and hardworking folks are merely in the wrong place at the wrong time.

Sadly, this is inevitable in an imperfect world with imperfect people who can't observe and evaluate everything that everyone is doing.

Please trust that most organizations try to eliminate bias and make this as fair as possible. There are many things you can do to make sure that, at least, you get a fair appraisal. In fact, by following the concepts in the previous chapter—truly understanding your organization's big picture, where you and your supervisor fit in, what is expected of you and how you and your supervisor are to be measured, you are already way ahead of the game.

One thing is universal—almost everyone feels they are equal to—or better than—what grades they receive on their performance appraisal. I remember walking the halls of Caltech and an associate of mine told me that every single professor could justify—in their mind—why they should someday win the Nobel Prize. Of course the egos in a University have no limits. But even outside the hallowed halls, we know from author Garrison Keillor that "all the children in Lake Wobegon, Minnesota, are above average." Perhaps it is human nature.

Like schools that grade on the curve, most organizations also establish a target distribution function of performance. Perhaps there are alphanumerical "grades" such as A, B, C, D, E or 5,4,3,2,1. Some organizations evaluate employees on a suite of categories including performance, output, impact, values, skills and competencies, teamwork, and promotion potential.

Given the practical realities of needing to distribute compensation, raises, bonuses, equity, and promotions, most larger organizations feel they must establish guidelines. For example, your company might prefer to distribute performance ratings along a curve such that approximately fifteen per cent are regarded as exceptional, twenty-five per cent are very successful, thirty-five per cent are successful, fifteen per cent are marginal, and ten per cent are branded in the least-effective category.

Maybe your company prefers to establish deciles or quartiles or even to force rank everyone at a certain level. Some shops are rigorous in keeping to guidelines and others are flexible. You should be able to ask your supervisor about this directly, and it makes a perfectly good topic of discussion during your job interview, *a la* "How are people evaluated or appraised here?"

Generally once each year at about the same time, typically around fiscal year-end, each manager reviews his employees' annual performance. The performance review calendar is set to give the company enough time to divvy out raises, bonuses, and equity based on how the company performed as a whole. A company that is going broke is not likely to give out bonuses or big raises to its employees. Often Boards of Directors need to review company performance *vs.* goals to establish the size of the bonus and equity pool.

Once the bonus, equity, and raise pools are established, each manager goes through the process of ranking and rating the individuals in her group including you. A good manager will examine several factors including your impact, effort, energy, creativity, teamwork, growth, and trajectory—a whole set of performance and values criteria. She will often give you a numeric grade which she will compare with others on her staff. For example, a score of 1A might be for someone who gets the job done in a big way, and ignites everyone around them with enthusiasm and care. A score of 1C in contrast, might be for a person who gets the job done, but does so by walking all over his coworkers and leaving behind a demoralized, defeated staff.

A good manager will think about you and the year you had, the degree of difficulty of the challenges you faced, the goals you set for yourself, and how you impacted her goals and the goals of the company as a whole. She will often check with colleagues for their thoughts about you, on occasion even using an anonymous three-hundred sixty degree evaluation where inputs from your complete sphere of coworkers are accumulated.

Hopefully, she will also ask for inputs from you before making up her mind. Here is where it pays to be concise and complete. I strongly recommend you keep a log of all of your accomplishments and impacts throughout the year. You are likely to forget many of the ways you added value—hosting an important visitor, making a personal sacrifice by working through the holidays to solve a product performance excursion, identifying and recruiting a key new employee, editing the company newsletter. Be as quantitative and financial as you can be regarding the impact. Developing a product that will sell "millions" is nice, but developing a product that sold $3.675 million in its first year is far more meaningful. And by all means, save these records which will eventually comprise your resume. After all, if you have trouble remembering all the good things you did, just think how hard it is for your supervisor.

Sometimes, you are only granted a few lines on a form to highlight your accomplishments. Sure—fill that out like a dutiful soldier, but write out in prose a full summary of what you did and how you grew the last year and append it to the form. Your supervisor may not read it or reject it, but she just might. And in any case, you will have a good summary of your impact for future reference. Don't let the bureaucrats run over your need to be fairly evaluated. And remember the adage in advertising, "Long copy sells."

It is likely that your supervisor will rank and rate all of her staff, and she will probably find that her distribution does not quite match the company guidelines. Her team may be top heavy as a result of her diligence in weeding out non-performers early, along with her ability to make stellar hires of new employees. She may be a brilliant teacher and motivator. She may be an easy grader or truly believe she has the best team in the company. Rarely do supervisors come in on the low side with their distribution. But as far as the company is concerned, she needs to come close to matching the overall distribution they established. She will then make the difficult puts and takes and finally come up with her list of employees' scores and rankings. If she can't do it, someone else will do it for her to her detriment.

One year when I was leading R&D, we really had a stellar year as a whole, and naturally, my distribution came in high. The rest of the company leaders, even acknowledging our impact, were resentful when my distribution was net above average.

On another occasion, one of my supervisors was extremely aggressive about weeding out weak performers while concurrently making stellar hires. His manager understood this and allowed for our team to net above average during the annual review process. But when my boss moved on and a reorganization brought our group into a new one, people who were truly above average were force-ranked into below-average scores, got bad raises and were resentful enough to quit.

After individual supervisors complete their team rankings and ratings, it is common that several managers at the same level will get together to collectively review each other's ratings to look for discrepancies and to formally integrate several groups into a master list. During these private and confidential discussions among managers, often key individuals are discussed. I cannot recommend enough that you make sure you are visible and well-known outside of your immediate department. A kind word or high praise from someone who isn't your manager carries a lot of weight during these discussions where it is generally assumed that most managers will be biased in favor of their own employees. Conversely, if you are unknown to anyone but your own supervisor, they will not be able to weigh in on your behalf and you might get moved down into the pack.

During these group performance integration discussions, it may come out that some bosses may be harder graders and others may be easier graders, and this is not fair. These discussions represent an opportunity to level-set scores and also to make sure that the larger organization is close to the target distributions.

Normally, people of approximately the same job levels are evaluated with respect to each other; naturally the higher the level, the higher the expectations for impact. Most companies try to balance the distribution at each level, rather than reserving all the high scores or low scores for a particular group.

As I mentioned, during the integration discussions, generally, managers tend to advocate for their people, and it is very hard for anyone to rate their own people as poor or deficient or unsatisfactory. No manager looks forward to those tough discussions with unhappy employees who received lower performance reviews, so it is far more difficult to populate the lower scores. And if an employee consistently gets low scores, he is often terminated which means that the manager has to have the most difficult of discussions followed by the challenge of finding a replacement. And if an organization has been rigorous in firing underperformers for several years, it is possible that low scores are forced onto pretty good performers. Nothing is more demoralizing for an employee than to fully achieve his objectives and still get a poor grade. Often in the scramble to populate the lower deciles, subtle and subjective differences push otherwise-good people down.

A poor and cowardly manager, rather than meet the distribution including high and low scores, may bring up the weaker performers to "average," and be forced to bring down the better performers to "average" in order to meet the target. This "spreading peanut butter" approach to performance reviews is very unfair to the folks who delivered more than their fair share, and they are just as angry as anyone who got a poor review would be. And moreover, these are the people you need to keep. If you lose a star because of an unfair review, it is one of the worst things that can happen to you as a manager. You should first ensure that your best employees know they are appreciated, and are treated accordingly.

In the book, *Winning*, Jack Welch describes how he almost quit GE after his first year, because he got the same grade and raise as a colleague who didn't accomplish nearly as much as Jack did. An eleventh-hour save by Jack's supervisor was probably the best thing that ever happened to

GE, as Jack would go on to raise GE's market cap by nearly 500 billion dollars at its peak.

Once the grades are reviewed and integrated for self-consistency among the groups and across the company, guidelines for compensation are applied and raises, bonuses, equity, and promotions are set.

Not every organization does it this way, but most do some variant of this. It is important that you understand your process thoroughly, in order to make sure you get fairly appraised and compensated.

Ways to Get Good Performance Reviews

To begin with, as we talked about in Chapter Seven, you need to deliver impact. If there is a good alignment between negotiated goals and impact, then meeting your goals is a convenient way of keeping score. We will talk about goals next, and maybe we did this backwards by discussing appraisals first, but I think you will find it helpful to understand appraisals and then learn how to use goals as a tool. And no matter how it is measured, making a strong, positive impact on the key measures of your boss and the organization is critical.

I urge you to make your value-creation as quantitative as possible and enumerated in dollars. You can create value by generating new sales and earnings, or by reducing costs. Cost avoidance is a great way to impact your company, and preventing losses—albeit more difficult to quantify and often underappreciated—is another way. As I mentioned before, I urge you to capture as many significant figures for your performance review and eventually, for your resume.

Next, again as we saw earlier from Yogi Berra, make sure you have looked for other ways that you have contributed to your team. It is extremely important that you keep a good log of these kinds of activities throughout the year as these are especially easy to forget. Almost always you and your boss will know what you did on the big stuff, but the little things often slide unless you make sure to track and remember them. They can be a key differentiator, especially in large, successful team efforts.

Finally, don't ignore your soft skills and soft impact that come from your values and personality. Try to be someone who is always energetic, enthusiastic, and optimistic. These are infectious behaviors and can inspire your whole team. In contrast, don't get in with the whiners that hang around the coffee pot and point out all the hypocrites in the front office or why the idiots in such-and-such department were causing all of your problems. Be someone who never wavers from the high road, who is always looking for a way to make things better and who brings out the best in all you encounter.

Values-Based Compensation Systems

I would like to touch briefly on organizations that purportedly measure values and competencies as part of their performance management systems. At GE, they called it "Make the Numbers, Have the Values." The idea was that certain managers could achieve their goals by single-mindedly driving their employees into the ground, pursuing their personal measures while sacrificing those of others, and leaving a lot of damage in their wake. These kinds of organizations and leaders are not sustainable as good employees leave for greener pastures, and organizations become dysfunctional instead of collaborative and complementary—or so the reasoning goes.

In general, I agree with the philosophy that values matter, though it too can become political and a popularity contest. A good, anonymous three-hundred-sixty-degree review can indicate where there might be problems—usually about people who put their personal agenda above others.

Are there people in your organization whom you really want to work with? What are some of their characteristics and how do you emulate them? They are genuinely nice, caring people who look for ways to help others. They lighten up a room with their presence. They energize their coworkers with their passion and enthusiasm. You will never hear them spread gossip or rumors. Their optimism is infectious. Make that person you!

Between having good values and consistently delivering meaningful impact to the organization, along with good, quantitative documentation of your contributions, you just need to have faith that you will get noticed and appreciated come time for performance reviews, compensation, retention, and promotion.

Chapter Nine

Goals

LOTS OF ORGANIZATIONS RUN ON GOALS—SALES goals, profit margin targets, productivity, free cash flow, new product vitality, quality, safety record, staffing, you name it. Often, Boards of Directors will negotiate with a company's leadership at the start of the fiscal year to establish goals that determine the overall size of the bonus pool for the company. If the company achieves these goals, the bonus will be at target and different individuals will get a certain bonus as a multiplier of their base salary. Exceeding, or falling short of targets will adjust the bonus multipliers as per the plan. Goals then propagate throughout the organization, presumably so that department and individual goals are in alignment with company goals.

As an example, suppose that a major company goal is to achieve a certain dollar amount of sales in a new industry with a new product line. Presumably the department charged with this goal will make this their top priority goal and a bunch of employees might have design and performance goals associated with the new product line.

In an ideal world, if you integrated everyone's goals, they ought to align well with their organization's overall goals, even if individual goals may vary considerably depending on their department and role. A company may have overall earnings, growth, and profitability goals.

Some individuals may be more focused on reducing costs, while others may emphasize growth. And together, they both need to meet their goals so that the company can meet its goals as a whole.

In practice, it is never quite so simple, but earnest leaders try very hard to create meaningful and measurable goals that pull their employees in the right direction.

Goals can be convenient when evaluating an individual's performance. If a salesman had a goal of selling one hundred million dollars of products in her territory, she either met, exceeded, or missed it. Salespeople are often measured and compensated based on meeting and exceeding their targets. A manufacturing engineer might be charged with achieving a certain production output and a certain yield or quality and at a certain target cost. These are pretty easy to measure.

If goals are well-established and align with the company mission, then goals offer a lot of value to the organization and are a fair way to base compensation. It doesn't always work out this way. Indeed the whole notion of goal-setting is fraught with issues and shortcomings.

Consider that some individuals sandbag and set low goals, and some managers are soft about setting tough goals and then holding their people accountable. Others set stretch goals that are outside of anyone's normal reach, but which often leads to higher achievement than setting soft goals. It is very difficult to cross-match the degree of difficulty of goals; for example, who knows *a priori* if inventing a new product will be more difficult than selling it? The world is so dynamic that a goal set fifteen months ago may no longer be relevant, but the employee might be held accountable for missing it simply because it was put in writing and it is easy to measure. Even worse is an employee who doggedly pursues a goal that she knows is long obsolete and irrelevant, simply because her compensation and job security depend on it being met.

In general, I favor measuring people on impact, while using goals as a tool. I know that the folks in Human Resources and the Legal department don't always see it this way, and it can also be confusing for employees. I hate to see an employee who met all her goals, but didn't make a difference to the organization and got a small raise or worse. Shame on everyone for setting meaningless or outdated goals.

In some cases, the situation may have made it impossible to achieve the goals. For example, suppose you had established a sales goal, but an unanticipated global recession happened and you and everyone else missed big. Is that your fault? Suppose you had a goal to jointly develop a new product with a key customer, but a quality excursion from another product line caused the customer to withdraw from any new activities with your company? You missed, but it wasn't because of your product design.

To a certain extent, good luck and good economies will cycle with bad luck and bad economies, and bonuses probably should rise and fall accordingly. Far too often, companies conveniently forget the good and bad times, and may discount someone meeting their goals this year because a vegetable could have sold out the product line in this economy, but during tough times, they still expect miracles. Other companies go soft during hard times and hard during good times, the net effect being that bonuses simply become an entitlement regardless of the results. Getting this part right is a challenge for most companies and is a big source of frustration for confused employees.

For example, suppose the economy took off like mad and your industry grew thirty per cent, but you only had a ten per cent sales growth target which you met and beat by ten per cent. You may have actually had a lousy year because you should have grown thirty per cent or more. Should you get the big bonus for exceeding goals? That isn't right, but unless you compensate on both the upside and downside, you are really causing dissatisfaction with your employees. On the other hand, there are always external reasons why goals weren't met, in which case, do you give a free pass for every time someone misses? Tough stuff, indeed.

If you are a supervisor, make sure you understand the rules and try to personally be consistent, but don't punish your employees out of principle if the rest of the company is behaving differently. Your key employees may just transfer to greener pastures next door.

Given the fluidity and interdependence of individual goals and the impact of external factors on your outcomes, I still like goals as a tool, but not as a final say in how an individual should be rated. That's why someone can miss all their goals and still have had a great year, and conversely someone can achieve all their goals and still have missed. It might not be as easy to explain, but it is still probably the most fair evaluation criteria.

I'll go back to what I mentioned about starting a new job, and recommend that, no matter how long you are in your current job, you should frequently touch base with your supervisor and your sphere of influence to make sure that the goals are still the right ones and have not drifted. If the goals do change, it is a good idea to capture the adjustments and send out a note to those that are impacted to make sure everyone is aligned and that there are no surprises at year-end.

SMART Goals

A very popular acronym for good goal setting is SMART standing for Specific, Measurable, Achievable, Relevant, and Timely. The reasoning is that a bad goal might be "I am going to lose weight" whereas a terrific SMART goal might be I'm going to lose fifteen pounds by June. I hope you see the obvious difference. Here setting, and writing down the goal will give you a greater likelihood of achieving it.

Stretch Goals

In most organizations, there is metric, i.e. "the numbers," that the goals seek to achieve, as we saw above. I mentioned stretch goals briefly—these are goals that are not intrinsically achievable—the "A' in SMART. Why would anyone set stretch goals if they were just setting themselves up for failure? There is considerable controversy over this topic. Some experts feel that this can frustrate and counter-motivate people while others suggest that setting audacious goals causes people to think outside of the box and they actually achieve a lot more than they would have otherwise from setting incremental goals.

I suspect that an organization that sets stretch goals as a rule needs to be a highly trusting one, lest employees fear they will be punished for falling short. I believe that certain, highly confident individuals, can set stretch goals, miss, and not feel badly. But that there are certain folks who will fret and worry over falling short and lose the self-confidence that is necessary to take on higher-risk/higher-return endeavors. You will need to decide how you fit in along this scale, and also to see what your company strategy and policy is here.

Many people are more highly motivated by setting several minor goals, and then get excited and feel satisfaction by checking off the ones that are completed. Where do you fit in this spectrum?

In most organizations, there seems to be a set dogma regarding goals and their impact on compensation. Your boss may be a soft-hearted manager who likes to set easy goals, or she may be a driver who sets tough ones. You may be tempted to set easy goals for yourself or exaggerate the perceived degree of difficulty and time needed to achieve them in order to ensure you get a top rating. Frankly, it is always better to avoid this petty gamesmanship if possible. You want to be someone who works diligently toward making the greatest possible impact on your company. In the end, you want to be measured on your impact toward achieving the main organizational goals. And you want to grow yourself and your coworkers along the way.

Relevant Goals

Early in my career, I made the mistake of setting and meeting tough goals that didn't really matter. I felt good about meeting them, but it didn't really do much for my career or company.

I mentioned in Chapter One that I interviewed at GE because a friend of mine was the campus recruiter. It was back in October of 1979, and I was in my fourth year of graduate school at Caltech. Not really knowing what I wanted to do after getting my PhD, I signed up to meet with several on-campus recruiters including GE. During the interview, I told my friend that I was planning on finishing in July of 1980 for an August first start. He told me it was a good company and I ought to go visit. Well come February of 1980, I spent two and a half days interviewing GE in Schenectady, NY, and then went to Kodak in Rochester for the next two and a half days.

I interviewed with three groups at GE—chemistry, polymers, and coal. I don't really remember too much about the interviews other than the coal guys saying "GE bought a big coal company, Utah International, so they want us to start a coal science group. We don't know what we are doing or why." I guess, neither did I, so why not give it a shot? I told them I would be available August first, reaffirming what I had told my campus recruiter the previous October. It turned out I was the only sucker they could get to take the job.

I worked very hard the next few months to finish my grad school experiments and to write up all my papers and thesis, and come hell or high water, I was going to finish on time and keep my August first commitment! And by the end of July, I was done.

On August one, I showed up to GE's Corporate Research and Development labs—on time as promised, bright-eyed and bushy-tailed and ready for duty. They were shocked. There was no office, no chair, no phone, nothing planned and nothing to do. Nobody expected me. They told me no one ever actually keeps the date they are supposed to graduate and start work. Typically new PhD's were six months late. I screwed up their budget on top of everything. Here I totally busted my hump to be on time, and it turned out to be a small negative.

Finally my new boss met with me and said that he still didn't know what we were doing, but that we ought to learn as much as we could about coal in case our new coal company ever needed help. I was to get involved with the top academics, and should set as a goal to publish an important paper in coal science in the coming six months. OK.

On February first, 1981, exactly six months from the date that I started, I submitted a paper to <u>Fuel</u> which was indeed published and was even noted in their sixty-year index the following year as "A Paper of Unusual Significance."

So far, I had met what I considered to be two important deadlines—starting by August one, and completing a paper for publication by February one. I met my goals. But . . . did anybody at GE care? Nope! I think my boss even forgot what he asked for. What gets points around GE is not meeting artificial deadlines, but making a difference to the company bottom line.

So the lesson I learned a wee bit too late in my career was that if you are going to set and meet hard goals for yourself, make sure that they are goals that really matter. Hence the "R" in SMART Goals is a reminder to makes yours relevant.

Years later, as a relatively new manager, I spoke to one of my most successful senior scientists about a potential new assignment. I actually forget what it was, but I recall him vehemently declining to work on the project, telling me that as a scientist, he got to work on maybe a half dozen major projects in his career, and he wanted to be certain that he chose them carefully to have the greatest impact. Once again, I learned an extremely valuable lesson from one of my employees.

Timely Goals

In a couple of the above examples, I met my goals on time and nobody cared. Usually, time is a critical element, and in a complex organization with intertwined and interdependent activities, being on time is hugely important. As we saw in Chapter Seven, doing the job well is a prerequisite for any kind of advancement, reward, and recognition, and often people fail because they are consistently late.

Customers have real deadlines. They will base their buying decisions on what is available when they need it. In the case of the semiconductor business, a launch date for a new type of chip is critical; often the profitability of a multibillion dollar production fab depends on a timely launch date which can give them first mover advantages and temporary price premiums until the competition catches up. Being on-time is especially critical in this industry where a fab might require an investment of several billion dollars, and obsolescence comes just a few years later. Truly, every day matters big.

There are several reasons why projects slip. Don't allow you to be one of them. Use the Pareto Principle to work efficiently and simply burn the midnight oil if you need to during a crunch. People in your organization will quickly learn that you are someone they can count on to deliver on time, and that will make you very special, indeed.

Personal Goals

In addition to the goals which affect your company's bottom line and your performance review, you ought to also set personal goals. You might include financial goals for your savings and investments, career goals for promotions, health and fitness goals, family goals, community goals, and goals for personal growth including reading and courses. As we will see, all of these personal goals will have a dramatic influence on your career. For example, someone who is financially secure may be willing to take on greater risk. And someone who is physically fit will have the energy to sustain intense work when needed, and won't suffer from prolonged injuries and illnesses that may keep him out of the office. Conversely, a leader who is out of shape, smokes, and is overweight may have trouble commanding the respect of his followers and suffer from low energy and high absenteeism. An individual who reads and studies will presumably learn and grow and become ready for bigger and better things. Don't rely solely on your company to set all your goals.

Summary

Whether your organization uses goals and goal-setting effectively or not, you will still benefit from setting goals for yourself that are truly meaningful. If you don't already know this to be a fact, you will need to take it on faith that real impact and achievement and personal growth have intrinsic rewards that far surpass any scores that a bureaucracy might impose, and that in the end, taking the high road by focusing on making a positive difference is always the best course. If you do not succeed because of politics and gamesmanship, this is a good organization to view from the rearview mirror as you move on to something better.

Chapter Ten

Building Your Personal Brand

COMPANIES SPEND A LOT OF TIME and money building their brands. What is a brand, and why should it be an important part of your career?

The dictionary refers to brand as "a kind or variety of something; distinguished by some distinctive characteristics to indicate kind, grade, make, ownership etc. What do you think of when you see the following brand names:

Toyota

Lexus

BMW

Ford

General Motors

Volvo

To me, *Toyota* evokes economy and reliability. *Lexus* means economical luxury. *BMW* to me means overpriced luxury with expensive maintenance and repairs. But that is my personal experience with BMW from a 7 series I just loved until the last year or so when a series of breakdowns cost me a king's ransom. I'm sure to many BMW owners and "wannabe's," BMW still means luxury styling and comfort, high performance, and prestige. Ford to me is an American automotive company past its prime, but who managed to not go bankrupt and is on the way back with economical and reliable transportation. And they make a great truck. GM is a company that was in chapter 11 and on the brink of being out of business, bloated by union and financial issues. But GM is also making a comeback by trying to regain preeminence as a technology provider. Volvo is a very safe yuppie station wagon driven by suburban soccer moms, and I wouldn't be caught dead in one!

What is your organization's brand? In my last company, the folks in Marketing worked very hard to create an image as "The trusted industry partner, providing high quality solutions with speed and superior cost of ownership."

You may have a strong perspective on your organization's brand and on a number of other companies' brands. Doubtless you and I would not perfectly match on those images. That's OK. Brands are all about perception and no two people share the exact same experiences. My overpriced luxury could be your dream machine. Your "cheap and tacky" could be my "frugal good value."

Now consider your personal brand. The bad news is you don't have a billion-dollar media budget to advertise yourself. The good news is that most of your audience is all around you and available for free. While you can't always control how often they see you, you control what they do see. And this can be quite important. Because how others perceive you will often determine many things important to you like your employment, compensation, promotions, and assignments. Often these assessments reflect your brand since others can never see you as completely as you see yourself.

When going about your business, do you appear energetic? Enthusiastic? Do you walk quickly with a spring in your step and a smile on your face? Or do you shuffle with your head down? Hey maybe you are just having a bad day, but for the guy who sees you only a half dozen times a year, you just damaged your brand.

Do you have an "elevator speech?" Imagine you and the CEO have just gotten on the elevator together, she introduces herself to you and says, "What do you do for us?" If you fill those thirty seconds with clarity of purpose and enthusiasm, believe me, you will get noticed. She will go back to the office, write your name down, and the next time she runs into your group executive, she will make a subtle enquiry . . . this is how careers are often made or broken.

Are you always on time to meetings? Are you attentive and prepared? Do you come in with a notebook to take notes? Where do you sit? How do you sit—alert and upright or slouched? How engaged are you? Are you fun and funny, quick-witted? Cynical and jaded? Naïve? How would others describe you?

When someone asks you to do something, do you jump at the opportunity, then over-deliver ahead of schedule? Or do you say you are too busy or have more important things to do?

How does your work and output become your signature? Are you bold and risk-taking or are you timid and risk-averse? How good are your judgments? Are you careful and thorough? Impulsive? Do you document well? Would someone describe you as fast or slow?

How articulate are you? Do you speak crisply and efficiently and are easily understood, or do you wander about the topic and never give straight answers? How strong is your vocabulary? Do you use profanity? Jargon? Do you mumble? Are your thoughts organized?

Are you shy and reserved or outgoing and friendly? Are you warm?

Are you trustworthy? This is usually the most important aspect of your brand. And while you may never actually lie, if you don't deliver on promises or commitments, people may suspect you are untrustworthy. And if you have a hidden agenda, it may be more transparent than you think. This is especially true if all of your 'unbiased' suggestions are self-serving.

How do you look? Sorry to say, but this matters—take it from an old, stocky, bald guy! I fight hard to keep my body mass index below the trouble line and it isn't easy. But people who have crossed that line may project an image of being someone who lacks self-control or discipline, and sadly, it is worse for women. If you look sloppy and disheveled, it is hard to not think that your work will be any different. Fair? Probably not but we are talking about perception here and most folks don't take the time to look too far beneath the surface.

Sadly, stereotyping is more common than we might hope in a free and open society. If you are a certain gender, age, build, race, religion, or member of any number of groups, there will be people who will make assumptions about you that may, or may not, be flattering. For example, someone might believe that "all Koreans are industrious," or "all Russians are unscrupulous,", "all Irish are drunks,", "all women will drop out of their jobs at some point to raise a family," or "all white men get a break." Fair? Of course not. But it is reality so you best be prepared to deal with it. There is a fascinating book entitled *Cultures and Organizations—Software of the Mind* by Hostede and Hostede that takes stereotyping to an advanced level by rating over a hundred different cultures along seven semi-quantitative scales involving rigidity, comfort with ambiguity, power distance, masculinity and femininity, and so on. So while you may be the exception, the Hostedes found that on average, Singaporeans, Japanese, and Germans are high on the rigidity scale, while Jamaicans tend to 'not worry . . . be happy, mon.'

If you are German, certain engineering organizations might ascribe to you personally, all of the great engineering branding that BMW, Mercedes, and Porsche have built over the years. Lucky you! If you are Jamaican, you might have to overcome certain barriers in order to land that critical design job.

Are there stereotypes for engineers and scientists? You betcha! Ever read a Dilbert cartoon? How many people out there believe that engineers are nerds, left-brained, antisocial and introverted? That could be a barrier to overcome should you desire to advance to significant leadership positions. There may be people within your organization who fear putting you in front of customers, investors, or the public just because of this stereotypical image. And frankly, how you perform in front of the public, investors, and customers is a major key in advancement to the most senior levels.

Obviously, there is nothing you can do about where you were born and who your parents were or even what preconceptions and stereotypes exist in your environment. But if you are aware of these stereotypes and they don't match who you are or where you are going, it will be important to change those images and perceptions through your actions, behaviors, and demeanor. If people think engineers are antisocial nerds, you can be warmer, friendlier, and more outgoing. If people perceive your culture as laid back, show them your drive. If people perceive you as rigid, bring out the flexible in you and show it proudly.

Perhaps you can discuss with your friends or supervisor how others see your brand. If it doesn't match who you really are, or if it isn't what you want it to be, then you need to go about changing it. Surprisingly, this can be easier and faster than you think. Walk a little faster. Speak a little louder. Always have a pen in your hand and a smile on your face.

And always, always be aware that negative press outweighs a positive brand by a long shot.

Lastly, you are also likely to be a member of several formal and *ad hoc* groups. Your company already has an image that will reflect upon you, and within your company, your department and section may also have unique characteristics. What kind of image comes to mind when I mention, for example, Goldman-Sachs, GE, IBM, the Navy Seals, the NY City Fire Department, Maytag, the US Postal Service, MIT, the US Congress? You might be a part of outside organizations such as the Lion's Club, the PTA, the Toastmasters, the Newcomers Association. Please keep in mind that you and your behaviors also affect the brand of everyone else on your team. If you work hard, communicate well, are honest and open, make good judgments and so on, you make all of your associates look good, and conversely. So please do! All of your pals are counting on you.

Chapter Eleven

Building Your Personal Stock

LAST CHAPTER, WE SAW THAT EACH of us has a personal brand. Now we will talk about our personal stock value.

So what do you think your stock is worth? I don't mean the shares you might own on the market; I mean your personal stock? Is it on the way up, down, or flat? Which way is your meter running these days?

By 'personal stock,' I mean how valued are you as a coworker and employee? How much trust, credibility, and influence do you have? How are you valued outside the company? How much would your loss be felt if you left?

Your personal stock is far more important than your performance review rating, which merely reflects a fraction of your total value to the organization. Your review is primarily about how you did *this past year*, and is subject to a whole bunch of outside influences such as the current state of your business segment and the degree of difficulty and risk in your work. Your personal stock is the sum total of all of your education, training, experience, contributions, and impact over the years, along with your perceived potential for the future, your values, your interpersonal skills, your marketability, and all the chits you built up and all those you cashed in.

Your personal stock is different depending on whom you ask. Your employees, coworkers, supervisors, other senior executives, and even competitors and recruiters will all have a different point of view. Sometimes these views vary wildly from person to person. Maybe you worked great within your organization, but you had a screw-up or bad conflict that disappointed someone outside of your organization.

Jack Welch, in *Winning* (p 282-288), talks about bosses who have to expend their political capital (i.e. their personal stock) on behalf of some of their employees. You really don't want your boss to have to do this. Welch talks about three of these situations in particular. The first includes value indiscretions. For example, if the company is on an austerity campaign and you insist on flying first class or arranging offsite meetings in plush digs, this could cost your boss some capital. The second is credibility—either lying which is rare, but perhaps telling half-truths or playing things so close to the chest that it appears you are an impostor. The third is "career lust"—wearing your career ambition on your sleeve, hogging meetings, self-promoting, seeking disproportionate credit for your role, gossiping about rumored reorganizations and so on. Not only does this behavior cause your stock to drop, but may also damage your supervisor's stock, who may be asked why he puts up with you and your antics.

Over the years, I have had to defend certain employees whom I valued, even though they caused problems. One individual was ambitious, outspoken, verbally compulsive, and may even have suffered from Asperger's Syndrome which is a high-functioning syndrome of autism. I am not qualified to make this kind of diagnosis, but many folks in the hard sciences are quite antisocial or socially unaware and clumsy, and seem almost incapable of experiencing empathy. In the extreme, they fit the Asperger profile. Some famous people associated with Asperger's include Bill Gates, Albert Einstein, Charles Darwin, Sir Isaac Newton, and Thomas Jefferson. The individual on my team was extremely competent, delivered effectively on his projects and with customers. But he damaged relationships across the company and I was forever defending him. Would you fire a Gates or Einstein, even if he was quirky? I made the hard call to keep him, but it always caused my personal stock to drop and I never looked forward to the annual discussions about him or with him.

Of course, you can add to your boss's stock as well. If you do something great that brings in a lot of money for the company, or makes your team win big, your stock goes way up and it also adds to your boss' capital. This is especially true if he had to stick his neck out to hire you, place you, or support your pet project. You need to never forget that you have the ability to make your boss's stock go up or down. If his stock is up, he is in a better position to help you and if his stock is down, he is in a worse position to help you. This is the great symbiosis between all employees and their supervisors.

Like your brand, your personal stock affects a lot of things including pay, bonus, upward mobility, and even the amount of autonomy and latitude you are given at work. If someone has a great track record for consistently delivering on important work with big impact—and all the time with great energy, enthusiasm, and values—they can pretty much write their own ticket without a lot of second-guessing. If they cause their boss all kinds of problems where he is constantly defending them, then they might expect a lot more scrutiny.

This is the real stock market. Now that you are aware of it, how will you behave differently? Certainly you should try to assess where you stand, but be wary of relying solely on your perspective—you actually have the most distorted view of anyone since you experienced one hundred per cent of your performance and behaviors, and are fully aware of your efforts, feelings, and motives. But you don't set your stock price— the marketplace does, and it can have a very different perspective. When appropriate, actively solicit the opinions of others whom you trust including your supervisor, and if possible, with anonymous three-hundred-sixty-degree appraisals. Try to get good at reading body language and the political tea leaves of your organization.

I would caution you against focusing too much on image; your substance and capabilities and contributions are far more important, and in time, you need to trust that these will align with your image provided that you are earnest, dedicated, and consistent. But still, if there are misperceptions about you, it is a good idea to be aware of them and nip them in the bud.

Chapter Twelve

Projects, Assignments–What Happens If You Miss?

IN CHAPTER SEVEN, WE TALKED ABOUT successfully getting your important work done along with finding new ways to help your team succeed. In chapters Eight, Ten, and Eleven we saw how these affect your reviews, personal brand, and stock, all of which contribute to your compensation and runway. Recognizing that there are politics and imagery in every organization, in the end, taking on mission-critical assignments and delivering on them is still an important key to your success. You may hear, "It's who you know that really counts," but who you know in the corner offices usually occurs because you got their attention by doing something important.

Do not fall into the trap that there is nothing you can do to get noticed and advanced.

If you are taking on seriously tough challenges with an element of risk, sooner or later you will fall short. If you aren't missing from time to time, you probably aren't stretching far enough. Do not let fear-of-failure allow you to fall into the mediocrity trap. In this chapter I want to explore this, often subtle and complex, issue—'what happens when we take on an important challenges and miss?'

Certainly there are consequences of failure and nobody is happy about it. After all, to succeed and win is so much more fun. Most organizations recognize this and may even pay lip-service to employees taking chances. This works fine in the abstract. But if you happen to be on the team that made the big and losing effort, you may be worried about your future. Should you be concerned?

Let's start by examining the reasons you may have missed your bold target, and how you had behaved during the course of the project.

In one extreme, consider that you had taken on a good, informed risk—an important and appropriate project that would have yielded rewards commensurate with success; after all, a big risk ought to yield a big reward if successful. And let's assume that you had worked hard, worked smart, tried a number of bold and innovative approaches to the problem, but for one reason or other, it just didn't pan out. And further, that you had worked well as a team, communicated your progress and difficulties throughout the project, never oversold your capabilities or understated the risks, remained enthused but did not generate false expectations that everything was going great just before killing the project. In this case, your management should go out of its way to protect and even reward you.

This actually happened to me on a risky new light-bulb project with GE. In this case, we succeeded technically in developing a bold new concept lamp, but in the eleventh hour our business ran into temporary trouble and could no longer afford the investment to commercialize the project. The project was cancelled though the principals on the team got big bonus checks and stock options. Similarly, when GE's acquisition of Honeywell fell through for unexpected political and antitrust reasons, the large team who worked on due diligence and integration still received CEO awards. This is the way it should be.

On the other hand, suppose the project was started with poor assumptions; was oversold; the execution was weak; the team was dysfunctional; poor results were hidden from management; and unrealistic promises were made. In this case, the key sources of the problem need to go. That's simple.

Usually, things are not so black-and-white. You probably contributed to the projects failing, whether it was doomed from the start because of poor planning, or due to poor project execution. But chances are, there is plenty of blame to go around. Be aware that finding scape-goats to blame is common. Sadly, rather than truly understanding and growing from failures, our culture is one that would seem to come up with quick and superficial reasons why a project failed, blame someone down the ladder and mete out punishment, then quickly sweep the whole thing under the rug.

I advise you to not be glib and quickly and easily dismiss failures—these are often your best learning opportunities. Make certain that you truly delve deeply into the root causes for the miss and not be satisfied with easy or superficial answers. Try to avoid making this a scapegoating exercise, but rather, focus on what you can do the next time to avoid the same pitfalls. It is amazing how even seasoned leaders often skip the biggest and best learning opportunities that failures afford, and just jump to a quick explanation and move on. It's as if examining the dead body will somehow become contagious.

In a really terrific article by Amy Edmonson in the April, 2011 *Harvard Business Review* entitled "The F Word," the author establishes a hierarchy for failure including "intelligent" failure. Clearly "deviance, inattention, and lack of ability" would not be reasons you would want to have associated with your name. But in a number of other circumstances, often outside of your control, you can achieve intelligent failure that grows you and your organization and leaves you both stronger.

What happens if your company really needed the revenue from the high risk project to make your numbers? The project, having failed, now likely means that your leadership is forced to take costs out of the company. In this case, they might or might not specifically look to this group for savings. Most organizations would probably not reinvest in a program that was unsuccessful. Would you buy a stock that continued to drop with no signs of a turnaround? Of course not.

Assuming the organization cuts its losses by terminating your project, it might look like the people on the high-risk team are being punished for taking a chance and not succeeding. After all, to the casual observer, these people were on a difficult, high-risk project; they missed; and the company is no longer investing in their project. So what happens to the participants? Since other projects may not be abounding with need and people are not necessarily interchangeable, some individuals may get placed on a new project that is not terribly suited to their interests and abilities. They may get placed with a team or supervisor that they don't really care for. Or in the extreme, they might even get laid off. To them and outsiders, it may appear they are being punished for taking a chance, even if this was not the intent.

What can you do about this? One approach would be to avoid high-risk projects, altogether. This is the wrong approach. The reality is that if you want to do more than just hide under the radar—to be a star in your business and field and aspire to greater rewards and recognition, fame and fortune, you must take on high risk projects and succeed from time to time. It is hard to get to the top anywhere doing merely routine activities, and it is even difficult for people to simply survive these days without stepping out and stepping up. Sorry but that is the way of the world.

What can you do to soften the blows on those infrequent or even frequent misses? You need to build up a track record of consistent high effort and achievement. Expand your higher risk activities gradually after building a foundation of smaller successes. That way, everyone will know that the miss is more of an anomaly and not the rule for you. You also can try to maintain a nice balanced portfolio of projects—some high risk, medium risk, and some sure things. That way when one goes south, others will sustain you.

You may feel like you have no say in your projects—that you just do the boss' or company's bidding. In some organizations this is the case, but as you strengthen your abilities and track record, you also need to be driving for more autonomy that will allow you more say in project selection. After all, if you are the person with the best ideas, you could always be working on the best projects—your own. Don't wait for someone else to give you assignments if you have better ideas.

If you happen to work for an organization where the boss comes up will all the ideas and projects, tells you exactly what to do, and then blames all failures on your poor execution, then leaving this team should be a high priority for you. You will never succeed here.

It is generally a good idea need to broaden your interests and capabilities and relationships across and outside of your organization such that when the inevitable restructuring of projects, organizations, and people occurs, you are versatile and visible enough such that you are sought after, and can quickly contribute to several ongoing or new activities of your choosing.

And remember, perhaps the most unsuccessful inventor of all times was Thomas Edison. His attitude was that he didn't fail a thousand times; he successfully identified a thousand things that didn't work, which moved him closer to the one that did work. If you can maintain this kind of attitude, success will find its way to you eventually.

Chapter Thirteen

Getting Promoted

SOONER OR LATER, YOU WILL HOPEFULLY be ready for a promotion. For many individuals, promotion is synonymous with management, though not necessarily. I have worked for a couple organizations where there was a promotion ladder for individual contributors—you know—Scientist I, II, III or Engineer, Principal Engineer, Engineering Associate, Fellow. Any professional organization could offer similar titles, compensation, and perquisites for their more senior and effective staff.

In other organizations, promotion means management, initially managing individual contributors including possibly your former peers, and then becoming a manager of managers. One might manage a small business unit, then a larger unit, then several units, leading to a General Manager role and finally to officer such as VP, SVP, CIO, CTO, CEO, and Chairman.

Organizational Development (OD) is a fascinating study in its own right, and is a major strategic activity of corporate human resource departments. leadership teams, and boards. Each organization does it just a little differently, but the basic principles are the same. It starts with someone who wasn't a supervisor before who now becomes one, with expanded roles and responsibilities especially involving accomplishing things through others. In the next chapter, we will talk about how to manage and lead, but for now, let's first talk about promotions in general, and how to get you one or more.

To begin with, are you geared toward management or just greater responsibility as an individual contributor? This is not a trick question. You can have a great life as either, and often people make the mistake of thinking they have to get into management to be successful. Some of the happiest and most successful people don't have—and don't want—anything to do with managing others. This list includes artists, writers, scientists, engineers, and various professionals including teachers, physicians, attorneys, accountants and so on.

Being Versus Doing a Job

So often we confuse what we think we want to be, with what we actually have to do once we are there. For example, ask yourself—'Do I want to be a researcher, or do I want to do research?' Do you want to be a doctor or do you want to practice medicine? Do you want to be a lawyer or do you want to practice law? Do you want to be a manager, or do you want to manage? Do you want to be an actor or do you want to act?

In some cultures and languages, these questions don't make any sense. You simply are what you do. But in our culture, there is—or at least there may be—a big difference. Lots of people want to be managers, but they don't really want to manage, for example. Who needs the headaches? Believe me, giving performance reviews is just about as much fun as getting them; now consider that when you are on the 'giving' side, you have ten or fifteen of them to do. Don't forget that your Human Resource Department wants everything thoroughly documented, as well. Love paperwork? Enjoy budgeting? Can't get enough meetings? Boy do I have a job for you! Don't get me wrong; there are lots of great things about doing a manager's job, and we will get to them in the next chapter. But most people don't really think about that when they seek the promotion; they just see ego, prestige, and dollar signs.

There are a number of prestige positions whose day-to-day life can actually be pretty unpleasant. To me, being a physician would be great—lots of money and status, life and death in your hands. We all know the adage about mothers who want their sons or daughters to become a doctor or to at least marry one. Getting into, paying for, and completing medical training notwithstanding, when I think about the actual practice of medicine these days—ugh! I don't think I would personally enjoy being around lots of sick and infirm people, though I suspect diagnosis and treatment would be interesting and the occasional life-saving success would be satisfying. And if that were all it was, sure it sounds like a pretty cool job for many. But—seeing so many patients for so little time, spending so much time doing paperwork and fighting with insurance companies and Medicare, working long hours and being on call during holidays and evenings and weekends, all the while fearing and facing malpractice and spending time in court . . . no thank you! I think being a doctor would be much better than the actual doing, thank you very much. Do you see the difference?

I suspect that being a university professor sounds pretty good, very prestigious, but the doing can also be gruesome. The professors I know are all pulled in several directions, chasing tenure, constantly preparing grant proposals and trying to raise money, writing papers, balancing teaching with research, spending hours on tendentious yet often meaningless university committees, and usually working very long hours for relatively low pay and with limited family/personal life. One might suspect that the situation improves with tenure, but it doesn't appear so, at least to me looking at it from the outside. Job security is nice, but the chase for grant money, top students, great discoveries and papers seems endless with the stakes ever-increasing.

Lots of people want to be CEO's, but ask them if they like traveling two-hundred days a year, kowtowing to investors and institutions and regulators, facing public scrutiny for corporate missteps and misdeeds, and dealing with every sniveling and whining complaint within and outside of their company twenty-four hours every day, while missing your anniversaries and your kids' birthdays year in and year out—not so pretty eh? Yes, the pay and prestige is good!

One of my favorite employees back at GE told me that some people considered management a position, and others considered it a job; he felt that I was among the latter group and he liked that. I didn't figure that management was anything intrinsically special—just a different job and I happened to be better at managing than researching and for him, it was the other way around. We got along great.

My philosophy on management has been that my role is to be more of a servant to my employees—helping them to get the right projects and resources, accessing rewards and recognition, and protecting them from politics and budgets so they could focus on what they were good at. I was never a directive boss, nor did I focus on the trappings of being the manager; I focused on the doing part of management. I truly enjoyed the role most of the time. But there were plenty of times when it was tough—firing and laying off people, fighting with colleagues over missed expectations and defending against finger-pointing, cutting our budgets and projects, reassigning unhappy employees, and even burying a few employees who passed away on the job.

For others, the "being" part of management is the key. They like the prestige, power, money, being on the inside of decision-making, the perks—irrespective of the actual work involved in managing. Some people go into management because they no longer want to do the work of those reporting to them, rather than because they intrinsically want to manage. If you don't truly want to serve, management will not fit you well. Look hard into yourself before you make the leap. And if possible, take on a player-coach role that lets you do both for a while lest you decide you want to go back to the bench.

Maintaining Perspective When Your Career Is Stalled

Lots of people get caught up in positions they might want, but jobs they hate or vice versa. Don't let yourself become one of them. It is easy to lose perspective in this matter. So many positions other than yours appear attractive to the casual observer, especially if you feel you are in a career rut with little mobility or advancement. It is extremely important to maintain the proper outlook when your wanderlust drive takes over. I have seen people walk away from terrific jobs for the wrong reasons and then spend years in regret.

For the moment, let's take it on face value that you are stuck in your job from an advancement standpoint, and maybe that your learning curve has slowed down. You still basically enjoy your work, have considerable autonomy and impact, a terrific and supportive supervisor, good compensation and a nice place to live for you and your family. This is a tough situation because it is mostly good. Still, it is possible that your lack of advancement continually eats away at you, and you end up overhauling your life as a result, only to regret it later.

I have seen it happen often. I once had an employee in exactly this situation, where the fellow really wanted to become a manager. As the VP, I felt he was an outstanding researcher who would struggle in the leader role. He indeed left the company, dragging his family along to a higher cost of living location, was quickly frustrated in his new assignment and his whole family was miserable.

So be careful about losing perspective and only thinking about what you want to be, but not what you want to do or what you already have. I've said this often already—if you are not happy in what you do, it will be hard to be successful at it, whatever it is. Then your chances of advancement really get curtailed.

How To Get Noticed

Assuming there is some sort of dual ladder for individual contributors or managers, you most likely want to get promoted. After all, there is usually more money, prestige, and autonomy involved. There is also more responsibility, accountability, and risk, but you are willing to accept that having confidence in your ability to succeed at the next level. Great! So how do we get you there?

As I mentioned in Chapter Four, simply spending enough time in your current role and being successful is not generally enough. This is certainly a minimum requirement and occasionally people get promoted for longevity. But more likely you need to show a propensity to do more—a capability to move into the next role and beyond.

Chances are that a significant element of a higher level job includes some form of leadership. How have you demonstrated your leadership?

I personally am a sucker for enthusiasm and initiative. A person who goes beyond the expectations of their current role, and takes the initiative in bringing innovations to the organization and assignment is someone I will notice, especially if they take on new roles with gusto. What are some of the ways you can step out in your current role? Even if your job is highly proscribed, there are always ways to stand out such as recruiting and interviewing new employees, running social or community events for your company, mentoring or being mentored, leading outreach projects with interns, local schools, hospitals or museums.

At one of my employers, the company allowed every employee who chose, to spend a day in a major community service event in lieu of work. One lower-level employee jumped at the opportunity to coordinate the major activity of hundreds of us overhauling a low-income community recreation center. The logistics and planning were overwhelming—making work assignments, arranging for tools, materials, food and drink, toilets, transportation, and publicity. The event came off brilliantly, and not surprisingly, the event-leader's career took off like a rocket after that. Talk about getting noticed!

Often in large meetings such as all-employee information sessions, organizational leaders may solicit questions or comments from the audience. Sadly, mostly people shy away from speaking up if they have an issue, not wanting to call attention to themselves or perhaps to some of the warts of the company. If you are sincere, asking a tough question or taking a tough, public stand can be a great way to gain visibility, but it can also lead to notoriety. Be careful that your leaders truly want outspoken, candid comments. Also be aware that many people may consider this to be grandstanding for the sole purpose of raising your visibility, and you might be resented as a result.

At one company where I worked, we had frequent all-employee meetings led by the senior leaders, with Q and A at the end of each talk. The same few people always seemed to speak up. Interestingly, everyone seemed to recognize that that one of the outspoken folks simply wanted to hear his own voice. Another was clearly out to try to "catch" the company leaders in inconsistencies like you might expect from a political journalist interviewing candidates. Yet another was obviously self-serving to try and get visibility to enhance his career. And finally, one was truly sincere and interested in bringing up meaningful issues and engaging the whole organization in finding better ways to work. If you can't be the latter, speak up at your peril. If you really have a burning issue, it is often better to approach the speaker after the session and ask for a few minutes of their private time, lest you risk embarrassing them and yourself.

Emotional Intelligence (EQ)

When I think about who among my employees would make a great leader, I look for people with high standards and great character and values whom others will happily look up to and follow. I look for intelligence, innate curiosity, drive, good communication skills including written, speaking and listening, and emotional intelligence. Daniel Goleman wrote in his treatise, *Working With Emotional Intelligence* (EQ), about a couple dozen major competencies such as self-confidence, trustworthiness, adaptability, initiative, optimism, commitment, understanding and developing others, communication and influence. Normally when I see someone's career de-rail, it is because of a shortcoming in one of these emotional competencies rather than simply failing to achieve a particular technical or financial outcome. Luckily, unlike IQ, EQ develops over time and expands with experience.

It is rare that anyone has all the bases covered, but some people just jump out as ripe for leadership and promotion while others have a long way to go to get there, especially in terms of leading people. While organizations often hire for technical expertise, some explicitly seek out individuals who have shown an early propensity for leadership, as well. GE has always been one such company, as Noel Tichy and Eli Cohen wrote about in *The Leadership Engine*. GE hires for leadership and nurtures leaders throughout their careers through formal training, an elaborate selection and leadership planning process known as 'Session C,' and through carefully chosen rotational and global assignments. It's little wonder that you've seen ex-GE guys all over the world as CEO's of public and private companies such as Boeing, Honeywell, Home Depot, Chrysler, 3M, Cooper Industries, Neilson, and dozens of other major organizations. Other great leadership training grounds include Pepsi, Procter and Gamble, IBM, McKinsey, and the various branches of the US Military.

The Pros and Cons of Promoting Experts

In my opinion, one of the big mistakes organizations make is to promote people into management primarily because they have the best skills as individual contributors. I understand that this is natural. After all, why would you promote someone who wasn't the best in their field? Wouldn't the folks who were better than the leader not want to follow someone less capable than they are?

Sadly, the skills needed to be a great scientist, tax accountant, attorney, production engineer and so on—are often the wrong ones for significant leadership. Bending a molecule or spreadsheet is certainly a skill and an art, but it has nothing to do with organizing a team, establishing meaningful goals, holding peoples' feet to the fire to meet a budget and schedule, or hiring, firing, inspiring and energizing a staff. While it is true that some of your most capable individuals are also the most adaptable and have a good chance of succeeding in very different roles, it is equally true that promoting the very best individual contributor can cause serious problems with the organization. Plus—why would you take your best scientist out of her role, just to make her a mediocre manager? That hurts your business twice.

Consider the situation where a great individual contributor has been promoted to his first management role because of his content-skills and talent. As a manager, this person will immediately garner the respect of his employees, which is a terrific foundation and head start on establishing strong leadership. However there is a natural tendency for the new manager to believe he has all the right answers, and it is equally natural that his employees will also tend to defer all their decisions to the expert. Ironically, this will completely defeat the purpose of management which is to accomplish things through others. It fails to develop the employees into leaders, themselves, and simply has them doing the managers bidding—expensive technicians, indeed. After a short while, the best individual contributors will leave because their creativity and autonomy and growth are all stifled by the expert-manager, In my opinion, it is often better to have an excellent professional leader who is a strong generalist leading an organization of experts who know their content better than their supervisor. This forces the supervisor to defer to the individual contributors' expertise and allows them to develop, while the leader in turn is then prone to take on a supportive role. But I completely understand why others might feel differently here.

General Strategies For Getting Your First Management Job

I think that getting your first management job is one of the toughest transitions to negotiate. After all, recognizing that managing people is different from managing projects, there is a high element of risk for everyone involved in the decision to promote you. Once you have successfully managed a team or two, it is a lot easier to move you along. How do you overcome this perceived risk in order to earn yourself the chance and to learn and practice the leadership skills you need to succeed in managing others?

The best thing you can do to prepare yourself for leading others is to actually lead others. I know this sounds dumb at first, but really there are plenty of ways to learn and practice the art of leadership long before you are formally promoted. No matter what your skills and interests are, it is likely that you can take on leadership roles informally. After all, leadership is more about organizing others toward achieving a goal, growing and mentoring and inspiring your associates, and influencing the directions you collectively take.

Consider hobbies, sports, community service, company events—all as opportunities to grow and demonstrate your talents. Have you ever put on a play for a theatre company? You can take on the leadership roles of producer or director. Are you an athlete? Maybe you can captain a team, organize a league, or coach youth. Charitable organizations are desperate for people to step up, especially for fundraising. Political campaigns are always looking for people.

In your company, are there *ad hoc* assignments that involve leadership? Mergers and acquisitions involve a combination of due diligence and integration. Can you volunteer for a step out role here? Does your company have any special events that need a coordinator? For example, one of my employers frequently took the Board of Directors to different company sites or on customer visits. Taking charge of the planning and logistics here is a great way to stand out. One time, I got to host the President of the United States and Speaker of the House for an hour visit—talk about a great opportunity.

If there are no good matches for your skills and growth that are already planned, maybe you can think of some ideas and take the initiative. How about a clean-up day or a recycling day? How about an office-supplies exchange? Do you have a library or website that needs some TLC? Take it on. Does your company have a holiday party or picnic? You can plan it. You will learn a great deal, and if successful, you will certainly gain visibility.

Through your regular job, you can also develop and exhibit your leadership skills. I mentioned a few of these before including planning, inspiring, influencing, and executing on projects. Next time your boss discusses a new project, why not volunteer to pull together a timeline and budget and staffing as an exercise with her looking over your shoulder? *Ad hoc* project leaders are often needed, and while this is only a temporary leadership role, it lets you get your feet wet in driving a project to completion without really becoming fully committed to management. One of my early leadership roles was as a project leader where I was both an individual contributor and manager. In many ways, I felt that the program manager role was the very best position in the company in that I got to stay close to the hands-on work and also had a taste of management. When I finally moved out of the lab completely as a section manager in one of our businesses, it was with some remorse and trepidation.

Less formally, you need to generate good, strong ideas about the direction your team should take, and you need to voice them persuasively in meetings and discussions. It is a bit of a tightrope to walk, but if you propose a particular direction for your team's project, you certainly ought to have good reasons and defend them persuasively, but without appearing stubborn or unwilling to listen, or refusing to compromise. Good leaders find ways to draw out several conflicting opinions without disrupting group dynamics and harmony. It is OK to stir the pot from time to time, but you must always end up as a team player who enthusiastically supports the decision of the team once decisions have been made. In fact, it is often when you disagree with the team's approach but still go out of your way to make sure your team is successful that you will achieve your greatest recognition. And it is OK to say afterward, "You know, I didn't agree with this approach at first, but now that I see how far we have come, I realize that this was an excellent decision." Leaders who are stubborn and unwilling to bend do not last very long.

Summary

We saw in this chapter that promotion may or may not include managing others, but it almost always suggests leadership of some sort. Before you jump into a career in management, be realistic about whether this is something you will love and be good at. If your career is stalled, is it because of your organization or is there something missing in you? Before you leap to a major upheaval in your life in pursuit of promotion, make sure that you are moving for the right reasons; often a lousy supervisor or bad strategy can get replaced rather quickly and unpredictably, and your career will be right back on track before you know it. As always, grow yourself in anticipation of your next assignment. If you want to move into management, read and study management. Put down those technical journals and pick up the Harvard Business Review; you become what you read and study and think about. Look for ways to get visible without being obvious or obnoxious about it. Your time will come.

Chapter Fourteen

Your First Management Job

MAKING THE TRANSITION FROM INDIVIDUAL CONTRIBUTOR to manager is tough. As we saw in the last chapter, it is tough to get the job. It is even more challenging to succeed in it. After all, now, you need to accomplish things through others instead of by yourself. Perhaps you have a vague idea how to do that based on observing people in similar positions including your last supervisor. But there is no guarantee that your last boss was any good at it. You should certainly try to learn a little something from each of your supervisors, even if it is "How not to be." But you should also take this time to learn from others, especially experts who have succeeded in leadership roles. I commend you for reading this book as just one of countless books on careers and management. After you have read a couple dozen, you may see some common themes, though there may be vast differences in approach, as well. In the end, you need to be yourself and accept that someone believes in you or they wouldn't have given you the assignment. And by all means, you need to deliver and meet or exceed expectations, especially early in your management career or it might turn out to be a short management career, indeed.

As you commence your management career, you will be closely watched by your employees, peers, and supervisors to see how you are doing and how you are evolving as a person. If you are leading former coworkers, they will be especially attuned to any changes they might see in you including social as well as professional behaviors. You might get ribbed for no longer being willing to go out with the boys on Friday night to toss down a few beers and complain about management, for example.

Make sure you completely subjugate your ego about the promotion, especially if some of your employees wanted the job, themselves. Try to avoid using your positional power to get things done. If you can't persuade your employees that your approach is a winner, maybe you first ought to rethink your approach rather than stubbornly forcing it on the team.

People will follow leaders who have good ideas, substance and character. If your people aren't willingly following you, forcing them to through positional power or fear is not likely to be a sustainable approach. Granted, most organizations are not democracies, but people like to participate in the planning of projects whose outcome affects their careers and compensation. Your employees will generally have good ideas, and will much more fully commit to plans that they had a part in creating.

On the other hand, there are bound to be times when you will need to break an impasse or even make a tough and unpopular decision. Strong leaders face these with aplomb, accepting that they have been called to make the tough calls as part of their assignment. Just make sure that everyone understands your thinking process, that you were fair and open, and you had to make a tough call based on the best available information at the time. Then move on and hope you were right. People do not expect perfection and will cut you slack, especially if you admit you are wrong from time to time and make a quick mid-course correction.

So how should you cope with the new manager role? Since the fundamental premise of professional management is to accomplish your tasks through others, let's think about the "others" to which we will dedicate the next several chapters. Certainly your employees—the people you manage—are a major key to your success, and we will discuss them at length in Chapter Fifteen. But they aren't the only people in your universe. In order for you to succeed, you will also need to engage your boss, your peers, the extended organization including other supervisors and support staff, your customers, your suppliers, and even the outside community. Think of yourself as the nexus, the glue that interconnects everyone needed to get the job done.

Moreover, though you have relied mostly upon your expertise up until now, the most dramatic transformation in your career is to now focus on your relationships. I will devote the next several chapters to these relationships and how they are unique.

Before jumping into the specifics, I can't emphasize enough that your integrity and character and reputation are treasures that must be sacrosanct. Never put yourself in a position where you take anything but the high road. Never break the law or behave badly. Assume that all of your words and actions will appear on the front page of the local paper. Always treat everyone with respect and courtesy and kindness. Be absolutely trustworthy. There is no statute of limitations on confidences and that includes everyone including your boss, your spouse, your shrink, your attorney, your clergy, and even distant friends whom you are sure would not even remotely leak your confidence. I don't care whom you trust or how much—when someone shares something private with you, it is not your choice to decide if, when, and how you will share the information. You will not . . . never, ever. Put it in the vault. This is one aspect of your reputation that people will truly respect.

Shortly after getting my first management job, a very good friend who became my employee shared, in confidence, an amazing new dream job he was going to take with another firm. I was sad to lose him, but it was such a great opportunity that I certainly wouldn't blame him for jumping at it, and I was actually quite happy for him. He asked me to keep it quiet, which I obviously did. A week or two later, he told me the story was BS—that he was just waiting to see how long it would take to leak out. Of course it didn't. I probably should have been peeved that he put me through this test, but I was so glad I passed it that I just shrugged it off. And of course, he told everyone about it afterward.

And finally, you may realize that management is not for you. People can be difficult, unloading their problems on you and zigging when you needed them to zag. You may face petty egos, discrimination, substance abuse, sexual harassment, and a myriad of ways for disharmony to creep into your organization. Who needs the headaches? Maybe not you. If that is the case, look deeply into yourself and recognize quickly that this is not what you signed up for, and go back to the role of individual contributor before it is too late. I've seen people leave first-management assignments several times, and it has almost always worked out for the best.

Chapter Fifteen

Relationships With Your Employees–Building Your Team

IT AMAZES ME HOW WRONG NEW supervisors, and even seasoned ones, get this part of the equation. Since management is all about accomplishing things through others, who are these others? Your sphere of influence and stakeholders is large, encompassing your entire organization and a number of outside groups. Some managers focus on their boss; others on their customers or even shareholders. My philosophy is that the foundation of your success is with your employees—your direct reports. Your employees are the ones doing the heavy lifting, and they should be precious to you. So many ambitious and inexperienced managers think that, to get promoted they need to be pulled up from above. In truth, you will be carried up by your team. Your employees can make you or break you. Once you realize that you work for them and not the other way around, things will quickly fall into place and stuff will start happening.

You will probably spend more time with them than anyone else in your sphere, and they will probably need you the most. So let's start by exploring these relationships first.

When you are a new, first-time manager, you should have a crystal clear idea what your employees want from management since you were very recently in their position, and you should also know what they weren't getting before you became their manager. In general, what most employees want is a clear mission, a reasonable set of expectations, and fair metrics. They want the tools, resources, and enough time to succeed. They want rewards, recognition, and your personal appreciation for their efforts. They want you to give them credit when things work out, and for you to shoulder the blame when things don't. They want compassion and understanding when they inevitably struggle. They want personal growth and a career path for themselves. They want your advocacy of them to your managers and to the rest of the organization. They want you to listen and to be fair. They will generally understand that you can't give them and get them everything, but they want to be darn sure that you will try hard on their behalf.

They don't really want you to tell them what to do though they understand that this often comes from outside of your purview; and they absolutely don't want you telling them how to do what they do. After all, they are the experts.

They want you to ensure harmony among the team and to have the hard discussions with teammates who aren't pulling their weight.

Can you do all that? Yes you can, and more.

You may be thinking that your manager doesn't behave that way. I agree—most managers don't.

Many managers tend to be bossy, directive, self-centered. They are worried about themselves and their supervisors breathing down their necks, and the things that they are measured on in order to get themselves promoted, to get themselves a bonus, or to avoid failure and all the fears that come with that. They probably tell their employees, in no uncertain terms, what they need to deliver; and when; and how.

So your boss is all over you like a wet rag, and as a manager yourself, you have been entrusted with a mission that you need your employees to deliver on. Great. Welcome to management! Instead of multiplying your boss's grief and passing it on to your team along with the fear of God, far better is for you to get your whole team together as equals with your role as facilitator and advocate, as orchestrator. You explain the mission and objectives to your team as told to you from above. There may be parameters you feel uncomfortable about such as the lack of time or resources, and it is OK to share your concerns with your team. But then, you need to say something like, "We are all in this together, and try as I could, I was unable to get us any more time, money, people, parts . . . , so we are stuck with trying to make it happen. How can we pull this off?" And then engage your employees—all of them—and listen. You don't need to be the one with all the ideas or even any of the ideas. Your people are the ones who know best what to do and how. Even if you are the content expert, bite your tongue. Do not be the first to jump in or you will stifle your employees' growth and engagement.

Listen to your group reflexively by repeating what you heard. Summarize frequently. Write what they say on the blackboard or whiteboard or flipcharts. Engage everyone. And as a group, with your facilitation, pull together a plan. It doesn't need to be perfect, but it needs to lead to assignments and action. And as your team gets the elements of the plan pulled together, slowly you will be building commitment and ownership for everyone. Explicitly test your plan frequently for buy—in and for gaps and holes. You don't want anyone walking out of the meeting unwilling to support the team and approach.

One method I have used with success is "Fist-or-Five." As each important decision point arises in a discussion, you ask everyone to hold up a hand with a fist or a certain number of fingers up to five. Five means you wildly support it, while a fist means you can't live with the decision and you need to go back to the drawing board. People who show fists are truly putting a halt and a restart to the plan. They need to explain very clearly why they can't live with the group decision and they need to offer better alternatives. If one person consistently shows a fist while most everyone else is on board, you probably need to get that person off the team if you can. If you see a lot of ones, this isn't a good sign for your plans, either, so you probably need to revisit.

The whole point of this exercise is to get commitment to action. You are not there to pull rank or make demands. You are a facilitator who wants to find the best possible solution, and to make sure everyone is on board. And you need to express support and confidence for the team's ultimate success.

Frequent areas of manager-employee conflicts in innovation organizations relate to resources and timing. Everyone always wants more time, equipment, and support staff. Here is where your expertise is actually its most useful. After all, if a Marketing or Sales guy was put in charge of an Engineering group (yes, it happens!) an engineer might say that a particular task would take a certain amount of time and require resources, and it would be hard for the Sales guy to argue. But assuming you were recently in the same role, you can say "Nuts" and get away with it.

Sadly, allocating scarce resources to individuals and their projects or sub-projects is a very contentious part of the role. I have found that the best approach is to listen to the conflicts without judgment, and ask questions about content. Speak reflectively with your employee by paraphrasing what you heard. For example, you might say something like, "If I understand what you are saying, you feel the redesign of the vopalizer in seven weeks is unrealistic." When the employee has completed his story about all of his concerns, you can then say, "I completely understand what you are saying, and you certainly are the expert. If I could get you more time and resources, I certainly would. But we have been given this job to do and only these resources to do it. So although I recognize that it is risky given the shortages of resources,

I am not quite ready to give up on the project at this point, because giving up now will get us all fired, anyway. So let's work together to see what we can do to maximize our chances of success, given what we do have at our disposal." And then sincerely work together to do just that. It will be terrific that you and your key employees are working together, not as adversaries, to try an work around obstacles and get the job done.

Once the program is underway, you will take on the role of making sure everyone keeps to his commitments. If people slip, and they will, you need to quickly and privately seek them out and have a discussion about why things are slipping. It will surely be described as someone else's fault, and you need to listen with empathy. But you also need to be firm that their component of the effort is critical and needs to be achieved along with everyone else's. And that you hold them accountable, regardless of the external situation. You have no time for finger-pointing, and simply want to get the program back on track. And you are willing to listen to suggestions on how, even to the point of bringing the group together to explore alternatives; but you are not willing to give up on the mission or target without a fight. Nor will you allow any of your employees to hand off accountability, regardless of whether he is responsible for the delay or setback. Remember that you are ultimately accountable, as team leader, and you need everyone to share the same commitment.

Any relevant information you hear from your supervisors should be shared. The old school management, where the manager hordes information from his staff in order to maintain power, is simply passé, awful, and foolish. Err on the side of complete candor and trust and full disclosure, but of course, never betray confidences or share private, personnel matters. Nobody likes to be blindsided by new information. Surely things change in an organization including mission and priorities, human and materiel resources. Your team may gripe but they will also understand as they are certainly used to changes coming out of the blue.

As Charles Darwin said, it is not the survival of the fittest but rather, survival of the most adaptable. Make that you and your team.

You need to sincerely like, respect, and trust your employees. You personally do not need to be the brightest nor the most creative nor even the most senior. In fact, if you are the brightest or most senior or most creative, be especially careful to not assert your talent lest it inhibits the contribution and development of others. People will defer to you if you let them, especially if you are the formal boss, and this does not generally lead to better outcomes. You may see quickly and clearly the direction to take the team. You must have confidence that as they wander around the desert, they too will reach the same conclusions that you have jumped to, and that having gone through the thought process, themselves, they will be more committed and capable this time and the next. If they really go down a black hole, you can always bring them back if you are dissatisfied with their conclusions or approach, but pull rank as a last resort.

In general, you need to make it clear exactly what is your role, and you need to set the tenor for the rest of the team. If possible, be the first one in and the last one out each day and work your tail off to set an example. Remember that the pack only goes as fast as the lead dog. Don't ask your employees to do anything you wouldn't, including some of the grunt work. Don't delegate all the hard work and keep the fun things for yourself.

Let me caution against something called counter-dependency, which is essentially the "Us-Against-The-World" approach to management. This works really well in small teams, and generates a lot of loyalty, but it is not sustainable and can become pathological for teams. Really a much better approach is to work on "macro-empathy" which is trying to understand the whole organization and your team's role in it. And that everyone else has a role that is just as important to them and just as important to the organization. After all, your team will certainly need to flange up with others to make the entire organization successful.

131

You want to encourage this—what Jack Welch at GE referred to as "boundarylessness."

In terms of the actual make-up of your team, chances are that you inherited your first set of employees, and neither you nor they had much say in the group's makeup. Be assured that under your new leadership your team will evolve rather quickly, either through behavioral changes in response to your style and approach, or through actual personnel changes resulting from hiring, transferring, and terminations. Try to not be impulsive regarding personnel changes, and give people second and third chances to meet your expectations. They are just as new to you, as you are to them. Often truly disgruntled employees will find a way out on their own, perhaps requiring a gentle shove or eventually a strong kick in the pants. We'll see in Chapter Twenty-six how and when to fire someone, and refer back to Chapter Three on how to hire the best possible people for your team. Try to make sure that you make talent acquisition and development a top priority.

By all means, don't make the mistake of waiting until you have an approved opening to start scouting for talent. The approvals to hire are often ephemeral with the door slamming shut at the slightest shift in the business or economy. When you get approval to hire, be ready to fill the opening quickly.

Assuming you have a good network (Chapter Twenty-one), you should be frequently tapping into it, asking all your pals for great people who may be looking to move on to greener pastures. Alumni associations, and even contingency search agencies are replete with talent. It is a good idea to be perusing dozens of resumes each week, and making a couple of "feeler" calls to potential candidates. Most candidates will be understanding if you are completely open with them from the outset. You just simply tell them the truth—that you have no openings now but expect one soon and you want to be able to move fast. You saw their resume and wanted to see if there is a potential fit.

Your Extended Team

I briefly touched on support staff. These folks comprise your extended team and are gold. Your extended team might include your IT staff, machine shop, maintenance, facilities, test and measurement, library staff, legal, administrative assistants, travel agents—all those unsung heroes that can help your team if they choose to. Again, the more you go out of your way to appreciate these folks, the better service your team will get. And it doesn't take much, as this group of employees is normally taken for granted at best, and more likely, catches grief from most everyone else. A box of donuts from time to time, an appreciation lunch, and sincere thank-you notes that are copied to their supervisors go a long way. You should think about this group as your extended staff, and remember that they too can either carry you to the top, or drag you down under the waves depending entirely on how they feel about you. A nice touch is to invite your support people to your group meetings every now and then, and to show them what you are doing and how they fit in. Try to get them to offer suggestions on how they can make things work even better.

In general, I feel that making management personal is a good thing. I got in the habit of sending everyone in my organization a birthday card, an idea I shamelessly stole from Ken Iverson's book, *Plain Talk*, about his leadership in transforming Nucor Steel. I always included a handwritten and personal note expressing appreciation or interest in them. It doesn't seem like much, but often when I visited my employees' offices and cubbyholes, I would see several years of birthday cards posted, so I guess it mattered to some of them. I always sincerely cared for my employees and their work and lives and dreams, families and vacations and health issues, and they knew it. You can't fake genuine care. And it is simply one of the best parts about being a manager—having so many people to care about. You may have heard the adage from Teddy Roosevelt, "People don't care how much you know until they know how much you care." People will cut you a lot of slack for your blunders if they feel your motives are pure.

I just never bought into the cold, impersonal style of management. Sure there are tough decisions that all managers have to make including cancelling projects, making reassignments, giving poor reviews, reprimands and terminations. But in my opinion, a solid personal foundation between the supervisor and the employee makes these discussions more meaningful and impactful, not less.

Chapter Sixteen

Relationships With Your Peers

In each organization, typically at any given management level, there will be a big boss and several managers reporting to the big boss. Assuming you are one of those managers, the other managers will be your peers. You will have close working relationships with those other managers and it is important that you get this part right. After all, someday you might become their manager or one of them might become your manager. Plus, assuming you are colleagues, you will have many things in common and your projects and resources may be closely intertwined.

For example, suppose your company is involved in the design or manufacture of some behemoth thing—maybe a car or electronic system, a medical device, a large software package, a building or system of some sort. Chances are that the organization is set up around various components that need to be co-designed and assembled. Your organization might design and develop component A which is joined with component B, C, D, E, and F and you need to work with managers B, C, D, E and F to make sure it all works together. It certainly behooves you all to communicate well and get along. After all, everyone needs to succeed in order for you to succeed. Never take the attitude that "I did my part, now you do yours." You all succeed or fail together. Virtually any part of the organization can bring the whole enterprise down. Got a great design but poor manufacturing or supply chain? Nice product but lousy patent protection? Terrific processes but labor unrest due to poor HR practices? You need to get it all right or you will get it all wrong in the end. So watch your peers' backs and hopefully they will watch yours as well. Curve balls can come out of anywhere.

What else might you do with the other managers? Chances are that the big boss will hold frequent staff meetings among his managers and you will all participate. You might be involved in resource allocation and scheduling discussions, perhaps personnel issues including promotions, transfers, raises and bonuses. You will probably spend a lot of time with this team and they may become your closest colleagues. After all, they have the same job you do, basically, so you have a lot in common. How do you behave?

There are many things that peer-relations have in common with employee-relations. You want to establish trust, warmth, and open communication. You want to listen with empathy. If you have a gripe, take it private and discuss things for understanding and without anger. Always assume that this is a good person with good intent. Where possible, look for ways to help that person succeed including sharing resources and being accommodating to their needs where possible.

You should not be a pushover, especially if one of your peers tries to take advantage of your good nature. But there is no need to start out distrusting or being defensive.

There are a couple of unwritten rules. You certainly should not try to poach your peers' best employees. If an employee comes to you from another organization anywhere within your company exploring or requesting a transfer, ask her if she has discussed it with her supervisor already. If not, don't continue the discussion without her supervisor getting involved. If she says she doesn't want her supervisor to know that she contacted you, it becomes quite a delicate matter. It is almost better to simply end the discussion here. You may consider speaking privately with her supervisor, pretending that the idea of her transferring was yours and that she has no idea that you two are talking, but this usually backfires. Almost all secrets eventually come out, and your trusting relationship with all of your peers is more important than getting a new employee who can add value to your team. In other words, if she won't speak with her supervisor candidly, you are probably best to let it drop. Stealing a colleague's employees is a serious transgression of trust.

Sometimes your boss may come to you and ask for candid feedback about a colleague. This is also delicate since you are being solicited by your boss who is expecting thoughtful opinions and judgment on your part, but you do not want to be known as someone who talks behind other's backs, especially someone in the same line toward promotion. It is hard to not appear self-serving in this situation. On the other hand, if there are some serious issues with your colleague that your boss is failing to address or is unaware of, having opened the door to be confidential and candid, you may have an obligation to your organization to speak freely. But far better is to speak with your peer first about issues before speaking with the boss.

Another delicate situation often occurs where one or more of your peers starts to discuss another peer in their absence, especially if they have issues with the missing peer, and they want to solicit your opinion. It is hard to not go down this potential rat-hole, because you may share the same issues and concerns, and you may be glad to see that someone else confirms or supports your observations. This is an opportunity to take the high road, by saying something like, "I am uncomfortable talking about Terry, since he is a peer and colleague." Or you might say something cute like, "My mother always told me that, if you can't say something nice about someone, to not say anything at all." You don't want to become a gossiper or back-stabber, though I understand that it is almost impossible to avoid these situations. Still, if you do talk about someone else, you should assume that the conversation and your opinions will be repeated and eventually become public. Now how do you feel about your image as a gossip and back-stabber? Conversely, if you always take the high road by refusing to get into these kinds of conversations, your peers will quickly realize that they can trust you to not talk about them behind their backs either, and this is good for your reputation.

I have a couple rules of thumb about confidentialities that are useful. If there is a safety risk, I will always speak up. And if I feel that someone is breaking the law or is about to, I will take action. I tell all my employees and colleagues that they can speak openly and candidly with me about any topic in complete confidence that I will not take action regarding personal issues, except for these two situations. For example, if one of my colleagues is a substance abuser or is potentially creating a hostile work environment, I would have no qualms in speaking up to my boss. Indeed I would not wait until I was solicited; these are serious matters of potentially severe consequences.

But what about the case where I simply feel that my peer is in the wrong job for his skills, knowledge, or personality? Should I voice my opinion? I think so, if done in the proper spirit, especially if the boss and I are sincerely looking to improve the organization and our outcomes. Again, this is delicate and needs to be backed up with examples of behaviors and actions that you personally disagreed with, and should never become a character assassination. This is not the time to bring up politics or petty disagreements with your colleagues. Focus on one major issue that gives you heartburn, and where it would be appropriate for your boss to know and to act if he agrees.

If you are not pretty darn certain that your boss is trustworthy, and if you would have serious issues if your discussion leaked out, it is probably best to keep your opinions to yourself. And if you are a supervisor who truly values candor from your employees, remember that just one slip of confidential information will destroy lines of communication forever. As nice as it might be to speak privately with one of your leaders about confidentially-obtained information, be aware that they will now assume you will share their confidential information as well. Of course, this does not apply to legal counsel where confidentiality is mandated.

If you do encounter a situation where you and a colleague have a disagreement that you can't resolve, it is perfectly OK to go right to that person and say something like, "You know, I value your judgment and capability in most matters, but on this particular situation, I disagree with your approach. And I plan to bring it up at today's staff meeting." Then do. Disagreements among trusted colleagues, including hearty debate, are the things that allow organizations to progress. There is no need to suppress sincere and substantive argument for the sake of harmony—at least not in American culture. Other cultures, for example Japanese business culture, will tend to make harmony paramount, with disagreements taking place privately in a highly stylized and nuanced manner. Westerners are more direct.

I can't emphasize enough that an atmosphere of trust and genuine teamwork and commonly held goals are necessary foundations for open debate to occur without damage. In highly political organizations, it is best to lay a bit lower and have less direct and less public disagreements. It is unfortunate that some supervisors require every encounter to have a winner and a loser, rather than finding a way for everyone to be a winner. If you are in this kind of organization, it is a good idea to consider an exit strategy rather than trying to work the system. Remember that all organizations have politics, to different extremes.

There are all manner of discreet discussions among peers, and you need to make sure these remain private, essentially forever. There is no statute of limitation on trust. Further, you need to be especially sensitive to keeping your colleagues apprised of things going on in your organization, or theirs, that they ought to know about. Public surprises are a serious source of embarrassment and are to be avoided.

If you do happen to flange up with your peers on projects or activities, go out of your way to make sure that they look good and that you give them more than their fair share of credit. Hogging the spotlight does not make you or your team look good, and can seriously damage trust and working relationships.

Always look for ways to help your colleagues in any number of ways including personal and professional. If you have a great employee, encourage her to transfer to your colleague's team if you think it is a good career move. Conversely, never dump your weak employees on to a buddy. If you truly feel that a change in organization will work for everyone, go ahead and transfer someone who has issues, but make sure that your colleague knows the "warts" and risks. Do not be relieved that you have unloaded a problem employee; this is not a time for *caveat emptor*—*i.e.* to let the buyer beware. You need full disclosure here.

Always be generous with your time and resources with your peers. If you find a good candidate, recommend him to your peers. Share articles, courses, and industry gossip that may be important. Look for ways to show that you are truly on the same team and not competing. If there is a shared resource in short supply, have a thorough and open discussion with your peer to see who has the greatest need for the common resource, and always make it available where it will have the biggest impact. Look for ways to share and compromise and don't be the one who always needs to get his way or be first in line. Your number one priority and his number one priority may not be the same, but you ought to be able to find common ground on the company's number one priority. If that doesn't work out, then at least seek solutions that are most fair to everyone.

Chapter Seventeen

Dealing With Your Boss

THERE ARE WHOLE BOOKS ON THIS topic insofar as the boss-employee relationship is so important and complex. In my opinion, the single greatest predictor of job-satisfaction is the individual's relationship with her immediate supervisor. That is why we have seen a sea-change in management philosophy over the last several years—moving away from the directive boss and more toward the supportive manager.

I had about fifteen bosses since I started working—it usually takes me a year or two to burn them out and drive them away, sometimes longer. Actually, in the case where you and your supervisor don't get on too well, you may be seriously considering to leave your organization; but before you act impulsively, remember that he is just as likely to leave soon enough if you can just lay low and ride it out.

Most likely, your supervisor has several direct reports in similar roles to yours, and his purview of responsibility is far broader than yours. If your organization has a particular scope, than his organization may include five or ten or fifteen organizations just like yours. He will not have the time to get to know you or your team or your issues to any degree of depth. Ideally, he is counting on you to execute your projects on time, under budget, with minimal intervention on his part.

Keep your communications with the boss crisp and to the point. Respect his limited time and attention. Show him that you can handle most of the day-to-day stuff and that you only call on him when you feel he needs to know something or can help.

Do not be shy about sharing bad news fast. You never want your boss to be surprised, even if dark clouds are merely looming as a possibility instead of already overhead. Often an early warning will give her time to prevent damage. If your supervisor is a kill-the-messenger type of person, this can be especially tough on you but take heart that this is still the best approach.

It is also likely that your boss can offer you help when you need it. It is far better to be pulling in the same direction, with you seeking his advice and support, rather than his taking on a role of critic and second-guesser. To the extent that you are able to handle your assignments independently and with aplomb, you will be rewarded with greater independence and authority. But don't be a cowboy and think you can handle everything on your own. It is not weakness on your part to leverage your boss on occasion. Just don't bother her with every little problem that comes your way.

Be careful to distinguish between a boss who wants a lot of information and who questions and challenges every decision you make *vs.* someone who is generally curious and interested in your projects, people, and progress. He may just be looking for ways to help. He may be challenging you as a way of testing you or helping you grow. Or he may be personally insecure and feels the need to micromanage everything, regardless of anything you may be particularly doing or not doing. I know it is hard to not personalize lots of management touch as an interference and distrust or disrespect, so try very hard to look at it in a positive light. You wouldn't want to be so independent that the boss just doesn't care about your or your team, would you? Also, let this serve as a reminder for how you ought to behave with your employees; they may not see your being helpful in a very positive light, either.

Sometimes a boss intervenes because things aren't going well, because he really doesn't trust or respect your judgment or approach, and he is genuinely worried that your screwing up will damage him and the organization. That's another manner altogether. Hopefully you can be candid and open and bring all the issues onto the table without fear. Ideally, you will bring forth your issues and problems, propose a few solutions, and ask your supervisor to weigh in on them. Just bringing up your problems for him to solve is not a good way to move up or earn anyone's respect. Be aware that asking for advice may commit you to taking it, so be careful what you ask for.

During a class I took from GE's renowned Management Development Institute, the Director, Steve Kerr, made a poignant observation about us highly-intense type—"A" GE managers. He told us that we go to great lengths to try to think of everything that could happen in our projects, pull together masterful slide decks on what we want to do and what we will need to get there, and then present the whole package to the corporate brass, almost daring them to find something wrong. The brass, in turn, in all their wisdom and experience, does exactly that—looking for holes in the arguments and taking potshots at the proposals. Kerr pointed out that this approach immediately puts you and the brass in adversarial roles. Far better is to identify the areas of risk and concern, be open and forthright with them, and make a plea for their help. Applying their wisdom and experience, together you forge a plan that will bring you success. In doing so, you flatter the brass, give them something useful to do, truly leverage their expertise, and get them on your side all in one fell swoop. Kerr is a brilliant guy, by the way, and has written a number of terrific books and papers on rewards, recognition, and leading organizations, including *Reward Systems: Does Yours Measure Up?*

Looking for ways to get your boss to add value is generally a good plan. Just pick your places.

<u>Lessons I Learned From Each of My Bosses</u>

With each new supervisor I had in nearly thirty+ years, I made several useful observations and usually learned one or two major lessons. I thought it would be fun to share them with you. I have mostly changed up the chronology to avoid embarrassing any of them.

My first boss was actually a shy diminutive professor who eventually won the Nobel Prize. From him, I learned that Yankees manager Leo Durocher was wrong—nice guys do indeed finish first. I marveled at the way Dr. Henry Taube would walk into the lab each day, and simply talk with each of his students one-on-one, giving his undivided attention. And then he would move on to the next. He showed, through his focused attention, that he sincerely cared about each of us, even if it was only for a minute or two to touch base. If your employees pop in and you continue to click away at your keyboard or Blackberry, what message is this sending?

Another boss taught me the importance of communication—through written and public presentation. Fair or unfair, people may evaluate you on the substance of your programs, but in reality they often make value judgments on how well you present them. The tacit presumption is that, if you present well, you must be smart, so your programs must be good. And conversely. He taught me that everyone is in sales of some sort.

Another boss was a hopeless pessimist. I swear he owned "end-of-the-world" insurance. Boy was it hard to work for him. I almost quit because I erroneously believed his constant rants that our situation was so dire. Your employees need to feel that, as their leader, you are optimistic; nobody follows a pessimist into battle. It was said of Joe Montana, the great quarterback of the perennial champion San Francisco 49ers of the 80's, that when he called the team into the huddle for each play, no matter what the score, he exuded the confidence that they would somehow find a way to win. And he did, often enough to make them believers.

One boss taught me about the importance of high personal energy. As Lee Iacocca says, "The pack only goes as fast as the lead dog." I know this may sound shallow, but sometimes, standing tall and walking a little faster, speaking a little louder and faster, can have a big impact on people's perception of you as a leader. And if you slouch, walk slowly with your head down, and mumble incoherently, what message does this send?

An astute supervisor taught me the importance of making a budget and keeping to it. Consistently keeping to my budget without complaint earned me so much credibility that when I did have a *bona fide* financial issue, I always got the support I needed. And frankly, making and saving money has always been an important part of any company's mission. Why not rally behind it instead of resisting it? Padding your budget says you care more about yourself than the overall good of the organization.

A very special leader taught me the importance of character and integrity. At the same time, his boss was a total sleaze. What a contrast! You can guess who I wanted to emulate.

A terrific boss taught me several important lessons. The first is to deliver the goods. Nobody cares about the size of your budget or how many hours you work as long as you hit your goals and beyond. No matter what else happens—make sure you deliver what is needed and expected, and more if you can. He also taught me that you need to get the very best people on your team, even if it means making tough decisions about average performers. And he was relentless and aggressive about upgrading talent. He also showed me the role of leader as advocate—standing up for one's organization during talent or budget or strategic reviews.

The next guy taught me about the love of science and technology. He was truly passionate about how things work and making them better and understanding them more. He was the scientist's scientist. One way to earn the respect of your employees is to genuinely care about what they do. An editor should love the language; a software team should be led by someone who gets software. Your team will feel it. You don't necessarily need to be the best at your subject matter (see Chapter Thirteen), but it helps if you sincerely value what your team is doing.

Another supervisor of mine was the most loyal to his people and the most caring. He was all about relationships and loyalty to a fault—maybe too loyal at times. He had trouble letting people go who were never going to succeed and in the long run, that helps nobody.

The next guy was not qualified for the job and was brought in because someone very high up in the company liked him. As a result, he got off to a rocky start with his team and never overcame it. I learned that a leader needs to establish his credibility before anyone will follow him, positional authority notwithstanding.

Even worse was the next boss on my list. He was also dropped in to the job utterly without relevant experience, and to compound matters, on day one he started telling people what to do and how to do it. He was very opinionated but without any basis, and he really got people mad—including virtually everyone across the organization. This was one guy who needed to read "*The First 90 Days*" by Michael Watkins. Since then, whenever I lacked relevant experience, I was very careful to listen to those who were the experts.

I also had a boss who was truly Machiavellian and ruthless—people said he would sell his mother for a promotion. He even went so far as to give me an assignment, then publically berated me for doing it! It was a set-up. I often learned more from my bad bosses than my good ones; always doing the opposite of this guy has served me well.

The next guy had almost the entire leadership package—except he didn't take good care of his health. He did not exactly have a good image, especially to those who didn't know him. I learned from him that you need to stay in shape for a lot of reasons, fair and unfair. Of all the challenges of leadership, maintaining my weight to below a body mass index of twenty-five has been one of the toughest. But as General Robert E. Lee said, "I cannot trust a man to control others who cannot control himself." With five of the last eight governors of Illinois in jail; with so many politicians cheating on their wives—Clinton, Schwarzenegger, Spitzer, Edwards, Weiner, Petreus—no wonder we have a country running out of control.

My next boss taught me about being politically aware of the undercurrents and hidden agendas that run throughout organizations, and about the importance of establishing close and trusting informal relationships throughout the organization, and the need for *quid pro quo*.

And with yet another boss, I have learned to always be candid with employees; and the importance of establishing a good core foundational strategy. Also, he sought to find ways to have fun while working.

You can see in my short list, that I encountered many different styles. In general, try to stick with the golden rule—treat others like you would like to be treated—and you won't go too far off.

How To Make Your Boss Love You

I read an interesting article in the September 21, 2009 *Smart Money* magazine by Ann Kadet—"Ten Ways to Make Your Boss Love You" who recommends that you focus on keeping your job by making your boss happy. This is always a good idea, though you hopefully know that I believe in making everyone happy—employees, colleagues, bosses, customers, suppliers and so on.

Here is a brief summary of what she wrote, and I will add a few more of my own.

1. <u>Put in the Hours When It Counts</u>. Bosses aren't really interested in how much time you spend on the job as long as you get the work done. But if there is a crunch, be the first one to put in the extra hours.

2. <u>Empathize</u>. If your boss "is being demanding, critical, stubborn or needy, chances are she's scared of failing or looking bad to her own boss." Especially valuable is to not only put yourself in the boss's shoes, but to take some action that relieves the stress or pressure.

3. <u>Show Initiative</u>. If you see something that needs to be done or improved, get the ball rolling. It is great to observe it and point it out, but far better is to come up with a plan for addressing the problem and get it moving.

4. <u>Be Positive</u>. Keep upbeat and optimistic. Remind your boss and everyone else all the reasons we should be enthusiastic about our future. Don't throw stones at issues but rally yourself and others behind opportunities.

5. "<u>Make Like Mini Me</u> . . ." People are most comfortable and trusting around people who reaffirm the validity of our own choices, says *Smart Money*. And they recommend you don't upstage the boss, either. Frankly—this one doesn't do anything for me. I prefer the diversity, but I can't speak for your current or future bosses.

6. <u>Get Your Face Time</u>. "Forced to choose between a competent jerk and a likable fool, any boss will swear he'd pick the jerk. But that is not how it works in real life," according to extensive studies. So get your face time and be more likeable, but not sycophantic.

7. <u>Defuse a Bad Situation</u>. According to extensive research, supervisors sort their employees into in-groups and out-groups, and once you are "out," bosses tend to not notice your accomplishments, but only your shortcomings—which

reinforces their impressions. The best way to break the cycle is a frank face-to-face discussion that you initiate.

8. <u>Ask For Help</u>. It shows humility and a desire to learn, but don't take this too far or you may appear lazy or incompetent. Save the boss for your toughest challenges.

9. <u>Connect The Boss's Way</u>. Deliver information to the boss in a way he can process it. And with the right level of detail. Some folks are micromanagers and others are big-picture people. Peter Drucker observed that some people are readers while others learn best through discussion. If you pile a lot of literature onto a listener, you may lose him. I had a colleague who needed to read everything . . . you could talk 'til you were blue in the face and it would do no good. His wife could call him six times about picking up the dry-cleaning and he would come home empty, but if she put a tiny post-it note on the refrigerator, he was perfect.

10. <u>Be a Conduit</u>. The old image of a boss who surrounds himself with "yes-men" who only agree and pass on good news is out. Best to be a trusting confidant who can bring important issues up quickly and in time to diffuse before they grow out of hand. Bosses prefer employees who speak openly and truthfully and who really have their hands on the pulse of the organization in ways the boss can't directly get. But don't be an alarmist, either. Not everything is a crisis so use your judgment.

These are from the magazine. I am sure there are more. I'll tell you what other things made me, in my role as a supervisor, value my reports:

- First—get the company good results. That helps me be your advocate and keeps my alligators at bay. By results, I mean progress, performance, savings, and sales.
- Help build great relationships cross functionally and with customers. I hate to have to use my capital to protect you when a colleague or customer comes to me complaining about something you said or did.

- Make good presentations and write well . . . it makes me think you are thorough, prepared, careful, and smart. And it reflects well on me and the rest of my organization when you come off crisply to the outside.

- Don't be self-serving—it is completely transparent. Always put the organization first and yourself last. If you are bragging, disrespecting a coworker, or pushing an agenda where it is all about you, this comes off really badly. Don't wear your ambition on your sleeve.

- Be impassioned and enthusiastic about your work. I am a real sucker for that.

- Always be honest and trustworthy. There is no gray area here. And honesty goes beyond telling the truth—it means coming forth with touchy information before I have to ask. It means believing in your data.

- Find a way to get along with your coworkers. Make an effort and then a bigger effort and don't stop there. I hate it when my employees are squabbling amongst themselves. I love a good argument about which is the best approach to an issue or opportunity, but I hate it when my leaders don't seek each other out or work well together. It looks petty and childish.

- Don't air our dirty linen or gossip unless there is a compelling reason to do so, and then try to let me know about it first. If one of my employees has an issue with a project or another person on our team, if they say negative things outside of our team, it often comes back to me and forces me to be defensive. You may have a legitimate concern but others outside may have difficulty putting it in perspective—such as knowing what is a hard issue or a minor one, for example, or believing that one individual's incompetence is a reflection on all of our employees. If word gets around that we are staffed with a bunch of lazy incompetents, how hard does it make it for me to ask for more time or resources, promotions, bonuses and raises for my team?

- Correct me when I am wrong, but please do it in private unless it is critical. Better to make sure in advance that I am completely informed and fully understand your issues and opportunities. Sending me a detailed e-mail is not sufficient. Make sure we talk through things and I get it right.
- Don't let others surprise me about an issue you are facing. It makes me look stupid and out of touch if I hear things first from outside.
- Call, text, tweet, and e-mail me immediately 24-7 if someone gets injured or gets into serious trouble.
- If there is an issue with ethics or legality—same thing. I must hear about it right away.
- Don't hang outside my door and socialize too long with my assistant. She has important things to do and her time needs to be respected as does yours. It is not fair to put her in the role of having to shoo you away. And the noise is very distracting to me. I prefer an open-door policy and that means I can hear everything outside of it.
- Do bring me articles and issues of interest related to the industry or business or management in general.
- Find a way to live within your budget. I will help you if you can't. I know there are times when unplanned expenses occur that can't be avoided.
- Do assume the best intent from me and your coworkers. You know that I am someone you can trust and am someone who always puts the best interest of the company first.
- Respect my time as I respect yours.
- Work safely and live safely.
- Set a good example for everyone.

Chapter Eighteen

Getting Along With Your Customers

In addition to employees, colleagues, and supervisors, you may get to deal with customers, suppliers, and even the public from time to time. Since you always represent your organization, it is important to know how to behave. Often your career will be accelerated or destroyed as a result of a brief outside encounter.

Customers are an especially important constituency. Almost always, you have customers, even if you are in the public or academic sector. For example, voters might be your customers; or people who employ your organization's services. Funding agencies and their contract monitors are your customers. Your students are your customers, along with their tuition-paying parents.

Some people are almost always internally-focused by design. Still they might regard their coworkers as the people whom they serve. By serving coworkers well, their real customers get better-served as a result. For example, maintenance staff or folks in the mailroom may not face the actual external customers who pay the bills, but they ought to serve their company's employees just as if they were the real customers. If they drop the ball, sooner or later the paying customers will feel it in cost, quality, and service. Do not feel you are any less important, just because you lack direct visibility to the folks paying the bills. Unfortunately, many senior supervisors and the rewards systems they put in place tend to forget this, lavishing praise and paying premium compensation to their customer-facing folks, while ignoring the folks in engineering, maintenance, logistics, operations, IT, and so on.

I always considered meeting with customers to be a special privilege since I was from R&D. While it is second nature for a salesperson, quality engineer, and customer service representative to interact directly with customers, for us techies this was reasonably unusual. After all, the folks in R&D tend to know too much and usually tell the truth, both of which can be kind of scary. Luckily the company rarely put us in front of an unsuspecting client without supervision from the salesperson.

I'm sure you all heard the old adage about two rules for dealing with customers: "Rule #1—The Customer is Always Right; and Rule #2—If in doubt, please refer to rule #1." Please take this to heart. It serves little purpose in arguing with a customer. You may actually be right; you may actually prove you were right; and you will still likely lose the customer. So who does that help?

Customers make our world go 'round. Without customers, there is nobody to pay the bills, nobody to serve, nobody to make things for, and nobody to design things for. Yes they are quite special and they know it. And they often behave that way. Sometimes you may be a punching bag for an irate customer, even if the damage was done years ago by somebody else who is long-gone. Customers have long memories and it is hard to overcome deeply-rooted animosity. Perhaps a decade ago, one of our customers employed our product, and we let him down with a serious quality excursion that caused him a lot of pain at the time. He may never forgive our company, even though virtually nobody from those days is still a part of our team today. For example, several years after we fired an incompetent sales representatives for a major customer, I got the pleasure of sitting through a half hour of being berated first in Korean and then in translation by this customer. Oh well. You should not be a total pushover, and need to stick to your guns by reminding the customer that, yes you understand that he was disappointed and rightly so, but that you and the new organization are different now, and you won't let these sorts of things happen again. But be careful about arguing over the facts as your customer sees them.

Of course, you need to be careful to not promise things you can't deliver to customers, nor to say things that contrast with what the sales rep may have told the customer before. And you should never talk price or terms with a customer unless that is your specific job. Be careful with over-confidence; if you make a customer think that solving their problem will be quick and easy, they may be unwilling to pay you value for it.

I know this sounds like customer meetings are a drag, but in fact they are just the opposite. You get first hand and direct information about what they want and need; the unvarnished and unfiltered truth. That is worth any amount of abuse you might get. And at times, it can be priceless. Let me give you a great example.

Back in the early 1990's, I was managing a silicone rubber R&D group. We had developed a liquid silicone that was being used to mold the dome sheets that make up computer keypads, where the intricate pattern of keys required a fast-flowing liquid to make sure that mold filled and cured quickly and perfectly. You probably know that keyboards don't contain an array of springs; rather there is a dome sheet of resilient silicone rubber, one dome for each key. It has to be perfect—you will feel it if the force on the @ is different than the force on the *, and you might just utter "$%%##@!" if the key sticks.

By 1996, our best customer for this technology was in deep trouble. Keyboard prices had dropped a whopping ninety per cent in just a year and our customer could no longer compete with low cost molded sheets from China. At wits end, our commercial team resorted to the unthinkable—asking R&D for help! They invited me to join the commercial brass—sales VP, marketing VP, regional sales manager, key account manager, for a visit to face the angry and frustrated voice of the customer.

When I walked into their plant, I was immediately struck with how quickly they were moving our product through production. I asked about throughput and determined that, each day, they were consuming virtually the entire daily production output from our batch mixer. And that they were only holding a couple weeks-worth of product in inventory. This is when I had an "aha" moment.

I was aware that we employed a very special and expensive raw material, and one that happened to be in very short supply, in order to give the product a one-year shelf—life. I reasoned that we could eliminate this expensive ingredient and the liquid keypad silicone rubber would still process well enough if they were willing to accept a shorter shelf life. So I proposed it to the customer as a cost savings and they loved it. It saved them millions of dollars a year, kept them in business for another three years before eventually succumbing to Chinese competition, and we retained a most important customer for our business. It was truly a win-win situation.

It would be unrealistic to expect our sales leaders to have known that there was a special additive that gives this product a long shelf—life and that we were also low on capacity. By sending the R&D guy to meet with the customer, this breakthrough connection became possible.

No matter what your special expertise is, you should certainly take advantage of meeting with customers if for no other reason than to make sure you truly understand how they feel about your products and services and to identify areas of improvement for them.

Getting the reputation as someone who interfaces well with customers can be a huge differentiator for your career. Conversely, if the field is reluctant to present you to a customer, you need to understand why and fix it fast. Take this responsibility very seriously. Often, you may be the only person your customer sees, and how you behave will strongly influence how he perceives your entire company. If you are polite, knowledgeable, thorough and prepared, he will tend to believe that about the rest of your organization. Obviously, there is a flip side to that perception.

It is especially gratifying to become personal friends with a customer. Make sure you don't take unfair advantage of this relationship or put your friend in awkward situations. Enjoy the advantage you already have with the customer without pressing it too hard. I can specifically recall one customer expressing appreciation for my not bringing up business during a social occasion. And if this friend happens to bring up an issue with your organization, even if it is well outside your purview, still take personal responsibility for resolving it. Your friend is counting on you.

Your customers may ask you for things that are inappropriate. They will pump you for information about your other customers who happen to be their competitors. Do not take that bait. If you leak competitive information, even public information, they will presume you will leak their information as well, and your trust—and information pipeline—will dry up.

Also, go easy on denigrating your competitors. Sure, you feel your products and services and quality and supply chain and costs are way better than the rest. Instead of trashing your competitors, which looks cheap and dirty, I find it is better to encourage your customers, with confidence but not arrogance, to test the competitive alternatives and decide for themselves. After all, they are going to anyway. If you don't think you have superior offerings, you are far better off focusing on improving yours than on trashing your competition because, like it or not, your customer will find out who has the best total package for them.

Be careful that your customer may be seeking the full gamut that you offer with a large multiplicity of features and benefits, and that he will only pick on the one or two deficiencies in your offering in order to get a better deal. It is fine to acknowledge your deficiencies, that you are working on improving them, and that you made necessary trade-offs in order to maximize the overall value to the customer. Most customers understand that controlling costs, for example in raw materials, may result in reduced performance. Never forget that your competitors have deficiencies, too.

Consider the times when you are buying a new car. You may have many requirements—initial price, maintenance cost, performance, reliability, safety, styling and so on. Do you quibble with the salesman because the highway miles-per-gallon efficiency is five per cent worse than an alternative you are considering in hopes that he will drop two hundred dollars from the price? A good salesman may note that you are choosing a higher-powered engine that necessarily has lower efficiency, but that they have an alternative that gives you plenty of acceleration and better gas-mileage. In other words, keep the big picture in mind.

And in general, you will always better-serve your customers if you try to keep in mind what being a customer feels like from their perspective. This should be easy since every one of us is a customer many times a day.

Our golden rule is "Do unto your customers what you would want your suppliers to do unto you." If you follow that, you can't go too far astray.

Chapter Nineteen

Relationships With Suppliers

SUPPLIERS ARE ALMOST SURELY A CRITICAL part of your business. Chances are that your suppliers provide raw materials or subsystems that you use to make final products, or software that streamlines your services and documents. Often suppliers are truly on your critical path for both quality and supply assurance. As we saw in the Fukushima tsunami and earthquake of 2011, a number of suppliers and sub-suppliers got knocked out, bringing a halt or at least a slowdown to a number of industries including automotive and electronics. While occurrences such as major natural disasters are rare, supply interruptions and quality excursions are not, and they can really bite you. Oftentimes, even with the best of intentions, your supplier will make a minor change that will wreck your products, and it can be a major detective effort to identify the root cause and implement permanent solutions. If you aren't close to your suppliers, this can become a near impossibility as your supplier often denies culpability and hides data. Far better is to have a trusting, committed partnership with your supplier so that together, you can get to the bottom of supply and quality issues quickly.

As I mentioned in the last chapter, as a supplier, you have a critical responsibility interacting with customers. Now with the roles reversed, often customers will disabuse their suppliers simply because they can. I don't recommend this. A good supplier is a precious asset and needs to be nurtured just as much as a customer. An open, trusting relationship can help avoid mismatches between needs and expectations, as we saw in the silicone rubber example, where we worked hard to meet a need—long product shelf life—that didn't matter to the customer.

It is rare that I ever fired someone for cause, but one of my employees was repeatedly abusive toward suppliers, and even went so far as to leak what I thought was supplier-proprietary information to a second source. Despite loud protests that this particular engineer was vital to our programs, I felt he crossed the line and had to let him go. I guess the temptation was too great to get back at all the customers who abused him by dumping on a supplier in the same way.

Yes suppliers are always nosing around to see how they can get an edge on their competition, or to try and value-price their goods and services. You would not likely want to share how much money a particular supplier saved you lest they jack up their price, for example. But generally, very open communication about your needs is best. You should never lie about or exaggerate your expected volumes in order to get their attention and a better price; this is just plain unethical.

Some customers fear that suppliers are out to steal their secrets and share them with your competition. As a supplier myself, I would never risk losing the trust and business of one of my key customers by sharing their secrets with another, though admittedly I have seen this happen. In Korea and Taiwan, for example, the semiconductor industry is especially incestuous, and suppliers are often a conduit for passing secrets between companies. Never allow yourself to be put in this position and don't do it to your suppliers, however tempting that might be.

As with your entire circle of influence—employees, peers, supervisors, customers, and suppliers, building and maintaining trust is your first priority; there is virtually nothing you can accomplish in an atmosphere of distrust. You may hold certain customers hostage if you have something truly proprietary, but in these days of technology moving at warp speed and intellectual property largely being ignored, your advantage of today will quickly evaporate and yet your poor relationship will linger forever.

An ideal customer-supplier relationship is open regarding requirements for quality, performance, cost, volume, and supply. Current and future needs should be communicated. Expectations regarding confidentiality should be put out on the table from the outset, including the potential for exclusivity for a limited period of time in exchange for joint development programs, data sharing, publications, intellectual property ownership, and partnerships. You certainly don't want to do all the heavy lifting with a supplier, just to have them hand it over *gratis* to your competition. But you also need to be reasonable about how long you should reap the sole benefits.

In general, you need to respect your supplier's need to grow profitably and be successful. You want them to be around for a long time, as well, and gouging them to the brink of bankruptcy is not in anyone's best interest. It is rare that you have an unlimited access to suppliers who can meet all your needs. And insofar as most organizations are both customers to some suppliers, and suppliers to others, you ought to both know how to look for win-win situations.

Chapter Twenty

Interacting With The Community and Public

LESS FREQUENTLY FOR MOST OF US, but still common and important, are personal and professional relationships with the community—your town or city, perhaps a professional society, maybe a university. The same rules apply as with your other spheres of influence—err on the side of candor; always be truthful and open. Recognize that these situations are opportunities to present your organization in a positive light because often, you are the only image of your organization to the community. The people you meet in these roles may never become customers or suppliers or employees and may never have anything to do your organization again. But they might, so treat them as you would any other important member of your sphere.

Often, a chance public encounter can have a big impact on your career. One of my bosses really got his career in high gear after just a few seconds on *60 Minutes* defending us in a frivolous lawsuit.

I regard all my outside interactions, at a minimum, as a recruiting trip and public relations vehicle. With universities, you just never know who might someday show up as a candidate for employment, or work at a customer or supplier, or perhaps even a regulator.

If you build up enough goodwill in your community, you may someday need to tap into it in case there is a problem—perhaps as a result of a spill or safety incident; or maybe a zoning issue. Nobody can predict when you will need help, but it always is good to maintain a warm and positive presence outside of your office walls.

I have always been enthused about corporate participation in charitable and community organizations such as the United Way or the local Chambers of Commerce. Charitable organizations represent terrific ways to hone your presentation and leadership skills.

It is a good idea to get to know your local elected officials including your representative to Congress, and key members of her staff. One time, my company had the great fortune of hosting the President of the United States and I believe some of my local visibility played a role in bringing him to our business. His visit gave our employees a boost and elevated our stature in the industry. And I got some great stories and memories along with a nicely framed and autographed photo for my office.

A crisis is a special situation that needs to be handled delicately. It is best to leave this to the CEO, Board Chair, Chief Counsel, and Investor Relations to determine the course of action and communication. If you are not one of these people, do not allow yourself to make statements regarding the crisis, other than, "That's above my pay grade," or "Our investor relations people will be commenting in the next hour or two." And leave it at that.

If you are a legitimate spokesperson for the organization, you probably already know how to behave—get out there fast or it will seem like a cover-up; give accurate and complete up-to-date information as you see it and leave room for uncertainty and changes; speak honestly with care and empathy; do not focus on you and your organization—focus on the community and those affected by the crisis, and do not appear that you are trying to offload blame or avoid lawsuits. Certainly one of the lasting memories from the BP oil spill in the Gulf of Mexico was of CEO Tony Hayward bemoaning the hassles the spill was creating in his own personal life. I hope you will know better than that if you ever get put in the public spotlight.

In general, it is important to be good citizens of your community and look for ways to add value and gain trust. Almost certainly, there will come a time when you will need friends from nearby, and all the things you have done to strengthen your community ties will come in handy.

Chapter Twenty-One

Networking

A VERY HUMAN BEHAVIOR IS TO seek out connections and join organizations. We are social beings. We meet a lot of nice and interesting people that way, and we enjoy a sense of belonging. Maybe you met your spouse through one of these groups, or found yourself a job or hired a great employee. Maybe you made numerous lasting friends or even an enemy or two.

Of course, the older you get and the longer you live, the more of these groups you tend to have accumulated. As I look back on my life, I can recall being a part of organizations like the Little League, Scouts, high school, college, grad school, my first employer at GE R&D, three jobs at various GE businesses. I have been members of two volunteer fire departments, one where I became a commissioner. I served on four corporate boards, the United Way board, and three university boards. I've lived in nine different houses and neighborhoods in six towns. In other words, I got to meet a bunch of different people from different walks of life.

Take stock of all the organizations and associations you have joined over the years. Have you stayed in touch with many people from them? Do you wish you had? It is not too late to reconnect now, and it is generally never too late to reconnect. One of the reasons that social networking sites such as Facebook or Linked-in are so popular is they satisfy our desire to make and sustain those connections. Re-establishing and maintaining old friends and associates is fun, and it is good for you and your career and business in general.

If you haven't done so already, you might consider joining one of these sites and seeing how many old friends you can find; you will be amazed at how your lives have diverged and perhaps re-converged. It may be a great big world out there, but it is also a small world in many ways.

A couple years ago, an old college pal tracked me down on Linked-in after a thirty-five year hiatus. We exchanged e-mails and phone calls, but try as we might, we had trouble crossing paths since we were both busy globetrotters. One early December day, I happened to be visiting our alma mater, snapped a quick picture and e-mailed it to him with the admonition that, if he was going to Chicago over Christmas, to not look me up since I was going to be in Aruba. He zipped back, "Which hotel?" Amazingly, we both holidayed just a few doors down from each other at the exact same Aruba resort and had a great week reconnecting; we are currently exploring joint ventures together in the coming years!

I have friends and colleagues scattered all over the world and across many industries. These represent a tremendously useful network to tap into if I need information, a job, a candidate, or a favor of some kind. But more importantly, I mostly focus on ways to look out for them. In a great book on networking, *Dig Your Well Before You Are Thirsty,* Harvey MacKay reminds us to give before we get.

Consider a situation where you haven't heard from someone for years and years, and out of the blue she calls up looking for help in a job search. You can't help but feeling a bit used. On the other hand, if she has been in touch intermittently—keeping you apprised of changes in her life and work and recommending candidates, customers, suppliers, or positions for you, then she will probably get your wholehearted support.

I carve out time each day to send brief notes to the people I have accumulated in my network—maybe for a birthday or to ask about their work or kids, or to send them an article of interest pertaining to their work or hobby. It only takes a moment, but if you are the recipient of some unsolicited attention, it has to make you feel good that someone out there is thinking about you and looking out for you. After all, the world can be so cold and impersonal.

A major networking opportunity is recommending and helping someone land a big job somewhere. Even if they are happy in their job and would not even consider a change, it has to be flattering that you thought of them for a big promotion, especially if it comes through one of the big executive search firms like Spencer-Stuart, Korn-Ferry, Heidrick and Struggles, Russell-Reynolds, Egon Zender, or Christian and Timbers. Getting on their short list is the fastest way up the ladder.

A lot of people in my contact list are executive recruiters from those agencies. We'll talk about them in Chapter Twenty-Three. It is important to give them names of truly outstanding candidates for their openings. If your list of top-shelf people is extensive, recruiters will contact you often and sooner or later, there may just be a job that you will want to bid on.

If you want a friend for life, help someone through a crisis—especially if they have recently lost their job. A lot of 'friends' disappear when you get fired, and this is the time when you need the most support and help you can get.

I would advise that you to not get too carried away—some people have virtual lives and friends through Facebook and never get out of their homes or offices. And others post foolish things about themselves, their coworkers, and their companies that can get them in trouble or even fired. Also, there is increasing concern about privacy. So do be cautious.

I cannot emphasize enough how important growing and nurturing your network is. As your skills and experience grows, so too will your network. When you step up to bigger jobs, your employer will be hiring your network as much as your skills and knowledge. Make sure it's up to the task.

Chapter Twenty-Two

Managing Managers

HAVING MASTERED THE BASICS OF MANAGEMENT and leadership, building up your skills, knowledge, accomplishments, and network, it is time to move up to managing managers. Chances are, this is pretty exclusive company with only a handful of opportunities in your company. Each organization may have numerous management positions—enough to try out several talented individuals—but they have much fewer posts at the next level and the competition is pretty steep here. After all, if each manager manages ten people, then there will only be one manager-of-managers for each ten managers. The pyramid narrows steeply, indeed. You may be discouraged about your future promotion opportunities, and find that it is necessary to go outside to move up to this level with any speed.

How do you get considered for this exclusive company? Consider what Jack Welch said in *Winning*. Jack made a few simple suggestions: "Do deliver sensational performance, far beyond expectations, and at every opportunity expand your job beyond its official boundaries." and "Don't make your boss use political capital to champion you."

Sensational performance is not too hard to understand. If you are leading new product development, deliver great products that have real performance advantages; and then help sell them; and then help the customer adopt them; and then transfer the learning to other teams and help them do the same. If you lead process engineering, develop great processes that produce those high performing products at high yield and great quality; find productivity including cheaper, better raw materials, streamlined processes, and transfer the learning to other teams and help them do the same. If you are on the commercial side, certainly deliver great numbers, grow sales, find new customers, increase margins, hire or develop fantastic people and move them on to bigger roles, communicate well, give credit to everyone but yourself, and come up with winning out-of-the-box thinking and strategies that can really move the company's sales and profits.

You get the idea. Whatever your mission, deliver and expand. I'm not a believer in sandbagging—you know, "under-promise and over-deliver." I'm all for taking on tough assignments, the kinds that most people shy away from, and then overcoming obstacles and still delivering. Sure it is risky and you might fail. That's actually OK if you learn and grow from your failures. Real leaders—the kind you want as your next boss—will know and understand this.

What about costing your boss political capital? What does this mean?

Obviously, if your organization misses a key goal, your boss will have to cover for you. This kind of goes along with the 'sensational performance' noted above, only in the opposite direction. If it happens a lot on your watch, it probably means that you aren't on a great trajectory. But there are other ways you can sabotage your career; what Jack refers to as being "a thorn in your organizations rear end." Often these involve corporate values and cultural transgressions. Maybe you spilled some confidential information to the wrong person. Perhaps the company is driving toward greater globalization or six sigma and you don't support the strategy. You should assume that anything you say will eventually come back to your boss or hers, and she will be making excuses or covering for you in some way. Maybe you have career lust and wear it on your sleeve, or worse—"insulting or disparaging the people around you in order to make your own candle burn brighter." Jack talks about "blaming others to cover up for your mistakes, hogging meetings, taking disproportionate credit for team success, and gossiping incessantly . . ."

Jack has a special place for people who don't exactly lie, but you have to pry the truth and full story out of them. What are they hiding?

Assuming you have paid your dues and passed the muster, there is a good chance you will be managing managers and at much higher level.

What's Different About Managing Managers?

Let's consider the major differences between managing individual contributors *vs.* managing managers.

To begin with, the scope of the job is far more encompassing. If you have ten direct reports, and each of them has ten direct reports, all of a sudden you are atop a hundred-person team with an annual spend of eight figures, and some pretty heavy expectations to deliver big impact to your organization.

Given the size and scope of the team, you can no longer get into the weeds of any issue—you will need to rely on your staff and theirs. You hope you have hired and coached well and can count on them to make sure that none of the details fall through the cracks. You become less tactical and more strategic. Your job becomes one of vision, communication, persuasion, and prioritization.

You will become more outwardly facing, with greater emphasis on relationships with your colleagues and customers and less on the actual work being done. You will need to become adept at distilling broad, complex issues and activities into concise summaries including status, issues, and plans. As before, you need to set a great example for your staff and organization, and you will be even more visible. The impact of making the right decisions is huge, but so will be your mistakes which will be magnified and scrutinized. Can you take the heat? You need to more be confident because you will be constantly second-guessed.

And yes—character, values, and trust are even more important. You are setting the tone for a large number of people.

You will interact with the organization's brass a great deal including the CEO, VP's, and the Board of Directors. They will be evaluating you, and by inference, your programs. You have a lot of mouths to feed, so the pressure can be great. The buck stops with you. Don't be someone who is constantly making excuses or blaming others. Accept that you are the responsible individual and take the lumps with the glory.

When starting a new job managing managers, you will certainly need to rely heavily on your staff, especially at first when you are learning the ropes. It is good to start with the assumption that your staff is comprised of capable experts, strong leaders, and people you can trust. After all, your predecessor placed them there for a reason. If you go in with an attitude of suspicion and disrespect, your folks might just live up to it. Especially sensitive are your very best leaders who may be nervous about proving themselves to you, who may be unhappy that you got the job instead of them, and who could be a flight risk. The last thing a new leader needs is a mass exodus of his staff; this will send a very bad message to your new boss. Your first order of business needs to be connecting with your staff and making them feel like they have a home and a future with you at the helm.

In time, have confidence that you will learn your new role and business, and eventually find out who the keepers are on your staff, and who you will want to move along. Do not be in a major hurry to make big moves, unless you have a serious issue with one of your reports in terms of style, or ethics. Rash moves will not earn you respect, and unless you want to lead by fear, which is an unsustainable approach, you are best to be thoughtful and deliberate.

It is likely that one or more of your new colleagues have special relationships with members of your staff. Perhaps they are close friends or arch enemies. They may be critical partners on a project. They might have historical *quid pro quo* arrangements in certain areas. Part of your getting the lay of the land in your new role will be to understand how your staff matches up with the rest of the organization, and if there is any good or bad blood. You may be solicited *sub rosa* by one of your colleagues to let you know that they need you to keep this person in their role, or to get rid of them quickly. It's a good idea to express appreciation for the information, but to let them know that you are not making any quick decisions or promises until you get a better lay of the land. In general, any promises your staff may have made before you arrived need to be kept, unless extenuating circumstances make this impossible. You need to make sure your team understands to let you know what is already on the books, and to not make any new commitments at first, before clearing them with you. In time, you will be more than glad to give them the latitude to make their own commitments.

Your new boss may have some biases about your staff that you should take very seriously. If you are replacing someone who was not successful, your new boss may blame your predecessor along with several members of his staff. Have the courage to tell her that you appreciate her insights, and that you want to make an assessment yourself. Buy yourself ninety days at least before making a move, unless your firing someone was a precondition of your promotion, or again if you are certain that this person is a disaster and their immediate departure will be applauded.

In time, of course you will want to establish your own new staff of trusted, respected leaders. Hopefully there are some good people from the previous administration and some new people. The people you hire as leaders will be a major reflection on you, so make these moves carefully. Replacing an individual contributor is painful enough, but putting in a new manager who flames out quickly can be devastating for you. After all, a bad leader can bring down his whole organization, including his boss. But, if you truly have made a mistake with one of your new leaders, despite the embarrassment, you must move quickly and decisively to remove him or the damage will fester.

As always, when hiring, look for competency, passion and enthusiasm, character and ethics, and give it the "smell" test—how will the organization regard this particular move? Is it bold? Are you taking a chance on someone fresh with a lot of runway but less experience? Are you putting in the obvious choice? Are you clearly and demonstrably upgrading?

In addition to adding and subtracting specific individuals on your staff, consider restructuring as a way to shake things up, put in your own team, and save face for many individuals. Often there are gray areas of responsibility which can be combined or split. If you feel your team is slow and bureaucratic, delayering is a great way to streamline a team and cut costs at the same time. Restructuring can be done to eliminate positions, and get people out of jobs where they weren't successful without the shame of being directly replaced. Again, as you explore restructuring, it is a good idea to confidentially test the concepts with your peers so that they won't be surprised or unhappy with the new set-up. Do not offer veto power to them, just ask for inputs. This is your show and you need to run it your way. It is perfectly OK to say something like, "Confidentially, I am considering revamping my organization to improve efficiency and shake things up to grow certain individuals. Here are some of my ideas. How will this affect you?" And listen and certainly take their thoughts into account. They might say something like, "Oh no, Harry would be a disaster in that role. A few years ago he crashed and burned when the last guy did something similar."

Summary

Moving up to managing managers is a big and challenging step. You will have much broader scope, and the potential for far greater impact. But the things that worked for you as an individual contributor and first-line manager will no longer work well at this level. Here you will be increasingly dependent on your staff, so selecting them well and delegating effectively will be critical. You will spend more time and effort working with your colleagues in the organization and with your extended sphere outside of your organization, and less time with your own team. If you try to micromanage and get into the weeds of every little project, you will quickly get swamped. Far better is to focus on the right people, processes, and strategy and let your leaders drive the work to completion. You will certainly be held accountable for the outcomes, and it is logical to be concerned when you have less direct touch. Try to avoid appearing uninterested or distrustful. Emphasize helping your staff to get the right projects and best available resources to be effective, and remember that always, you should put the interests of the entire organization first.

You will often read the business literature that talks about the difference between leadership and management. As you take on managing managers, you shift from management, which is doing things right, to leadership, which is doing the right things.

Moving the needle on a bigger organization is a lot tougher than for a smaller entity. It certainly takes great ideas and strategies which you should absorb from any source—your employees, colleagues, customers, competitors, suppliers, or the literature. If you solely rely on your own ideas, you will be severely limited. But good ideas are not nearly enough. You will also need to influence your team to take action. If you rely on positional authority, your strategies are less likely to stick compared to persuasion. Persuasion begins with character. People will follow leaders with vision and character and you need both. You will need to be enthusiastic, impassioned, and you will need to be a good salesman no matter what the title says on your business card. You are spending millions of dollars of someone else's money and they have a right to know where and why. You are investing hundreds to thousands of hours of your employees' lives, and they too have a right to know where and why. The ability to create and sell your vision across the entire organization is a major factor in your success, so take it seriously. Be patient as others will not necessarily arrive at your conclusions as fast as you have. Be persistent and committed, and accept that you may not get a hundred per cent of what you want at first, and that you may have to accept milestones and checkpoints along the way. That's OK. Just never give up on where you believe the organization needs to go, because if you as the leader blink, others will quickly lose their commitment as well.

Tough stuff? Indeed. That's why it gets lonely as you become a manager of managers. But here is where mountains get moved and real change gets driven.

Chapter Twenty-Three

Executive Jobs, Executive Search

IN THIS CHAPTER, WE'LL TAKE A look at executives—especially corporate officers. First, we'll see what it takes to do these jobs well; then we'll discuss how to get the nod, usually from the CEO and perhaps the Board.

Why You Want To Be an Executive

There are a lot of great things that executives enjoy. If you are intrinsically driven, you will enjoy the complexity of the issues and the development of strategies, the dynamics of human behavior and the joy of bringing out the best in people, the ability to influence, in a bigger way, your organization and beyond. If you are extrinsically motivated, you will certainly enjoy the prestige, the compensation, and the perquisites.

Generally, once people determine that they belong on the management track, moving on and up tends to come with the territory, though not necessarily exclusively. For example, the Group Leader role in research, engineering, or operations can be a fine place to spend a career—still doing science or engineering, while leading a small team as well. A sales manager may just love that customer interaction and may prefer to stay closer to the customer and far from headquarters and all the politics that go along with that. Perhaps a school principal would want nothing to do with life away from the kids and would loathe the politics in the district office. So continuously moving up and on is not axiomatic for all.

What Makes a Good Executive?

Do you ever look at folks on your Leadership Team, or executives elsewhere, and wonder if you have that leadership capability within you? Cathy Anteraisian, Gerhard Resch-Fingerlos, and Robert Stark, principals at the world's leading search firm, Spencer-Stuart, have written an insightful article on how they think about *Understanding Executive Potential*. Let me briefly excerpt what they said and make a few additions of my own.

The first building block is **Exceptional Business Judgment**. Questions to consider include 'Do I trust this person's judgment in complex, ambiguous situations? Has their decision making been tested when leading a team outside their area of expertise and in situations of great complexity and ambiguity'?

The second building block is the **Ability to Recognize Interpersonal Dynamics and Apply Them in Decision Making**. Questions to consider include 'How effectively does the executive read and respond to interpersonal dynamics in sensitive, high-stakes and complex situations? Does the individual understand the power of his words and actions on others and quickly create alignment among stakeholders with divergent interests? Can he or she successfully navigate politicized situations where personal relationships and a cooperative style aren't sufficient?'

Next is **Highly Effective People Management and Team Building**. Here you might wonder 'Does this person have a track record of building high-performance teams? Is he or she willing to hold people accountable when they fail to meet objectives? Does this person create an environment where people feel motivated to contribute, while also holding others to high standards'? Is she genuine? Does she put others' needs above her own?

Does this individual exhibit **Humility and Substance**? They ask: 'Does the individual show the mental flexibility to quickly evolve their thinking based on others inputs? How does he or she react to criticism of their ideas? Does he or she really listen to substantive input from people who know? Does he or she seek it out?

A strong executive has demonstrated themselves to be **Great Developers of People**. You might ask 'Is talent development a priority for this executive? How has he or she demonstrated this is a priority? Are there a number of individuals in the organization whose careers have been shaped through their relationship with this executive'?

Do they have **The Ability To Drive Change**? Questions to ponder include 'What are this person's strengths? Does he come up with the big ideas? Are they most skilled at executing an idea from somewhere else? In past situations of change, what was the individual's role in developing the vision, influencing and motivating others to embrace the idea, and driving to a result'?

That was the Spencer Stuart line-up, and who would disagree with the best and brightest in the field? I would add two more explicit topics of my own:

Does this person exhibit **Good Values and Character**? Questions I might add include: 'Does he or she maintain confidences, always and forever, across all business and personal relations? Does he or she deliver on promises and commitments? Does he sincerely care about his colleagues, employees, customers, suppliers? Is she unbiased and does she sincerely value diversity? Does she speak candidly and have tough conversations when needed? Does she have courage? Integrity?

Consistency? Proactivity? So much criticism of our so-called leaders in Washington and throughout our statehouses—on both sides of the aisle—is about their lack of courage in addressing obvious and systemic issues in financing government expenditures. They see the oncoming train, but because of political expediency, they choose to defer the hard and unpopular decisions to the future, knowing full well that time will only exacerbate the problems. They are self-serving cowards and the public knows it. Corporate leaders need to also have courage—to cut costs when needed, but also to invest into uncertainty. No financial model will ever tell you that an investment in technology or acquisition will pay off. It takes courage and conviction to take a stand and drive action, especially when knowing that the outcome is uncertain.

Does this individual **Communicate Effectively**? I would ask 'What is his private and public persona? Does he listen and speak clearly? Does he write crisply and speak with authority? Does she understand the power of stories? Can she read an audience and speak credibly and persuasively? Is he able to communicate across cultures and generations? Does she adjust her content and style to be appropriate for a range of audiences including board, other executives, technical experts, exempt and nonexempt employees, customers, suppliers, students, the lay public, and across cultures?

How do you score on these qualities? What is your biggest strength? Where is your Achilles' Heel? Must you be good or excellent in all these categories? How do your leaders stack up under these screens? What are some other areas you might look for?

Chances are, like most people, you have several outstanding characteristics, and a few areas that need work. It is rare that an individual has the entire package. Keep a close watch for areas you need to work on. If your company offers you career-coaching, you should take advantage of the opportunity for someone to speak candidly with you about issues and opportunities.

In my experience, you need one or two outstanding characteristics—great speaking ability or strong innovation or great intuition or tremendous customer engagement—to get noticed and onto the fast track to the C-suite. And that you also can't have really any severe Achilles Heel's. By that, I mean a glaring deficiency in one of the areas above. For example, a company would be very hard-pressed to promote to officer someone who has serious communication issues, has ethical lapses, suffers from cultural bias, or lacks interpersonal skills. Assuming you are OK across the board and have a couple stellar characteristics, you just might get noticed.

Some organizations are especially strong at developing leaders including GE, IBM, Pepsi, Procter and Gamble, and the US Military Academies. These organizations tend to promote from within, and with a surfeit of talent, often lose great, perhaps impatient, people to outside opportunities. Other companies have neither the leadership pipeline, scale, training, nor confidence in their own people to promote extensively. It is hard to cut your teeth at an organization and grow without making mistakes and doing damage along the way. I refer to this as "having warts." When someone comes along from the outside, they often look great compared to the internal candidate, simply because they can hide their warts pretty well during an interview. Obviously the internal candidate has screwed up and crossed swords with some of the incumbent leaders; if not, then she is probably more of a politician than leader, and I would be suspicious of her capability. I want someone who can go toe to toe with the other leaders on important issues of strategy and resource allocation.

That being the case, it is little wonder that organizations often go outside to add to their leadership staff. Most of their internal candidates have issues, and the rest have already left for better opportunities elsewhere. This sad fact of life brings into play a very important segment of the career game—executive search, aka headhunters.

Executive Search

I don't much care for the moniker "headhunter." It sounds demeaning, like they are sneaky poachers of talent. In fact, they are more matchmakers than they are corporate felons. And they need to become your best friend or you probably won't be going anywhere soon.

There are two kinds of search firms—contingency and retained. If one happens to contact you, it is OK to ask which they are, and if they are contingency firms, they are unlikely to have the plum jobs. Contingency searches are fine for technical and middle management jobs, but the retained firms really are the ones you want on your side. The biggest include Spencer-Stuart, Heidrick and Struggles, Russell Reynolds, Egon Zender, Christian and Timbers, and Korn-Ferry. Also, each industry has specialty boutique firms that may emphasize biotechnology, pharma, consumer goods, education, military, energy, finance, R&D, automotive, electronics and so on. For example, I work closely with Leadership Capital Advisors, a search firm in Chicago that is very strong in business, technology, and engineering leadership for the chemistry and materials industries. You should find out who the specialists are for your field in case they call.

Executive search firms are hired on retainer by the company they are representing. Typically executive recruiters are paid a fee around a third of the annual cash compensation for the job they are filling, plus expenses. They generally expect the search to last months, though some high profile CEO replacements may need to move very quickly.

Once the executive search firm has the contract with the hiring company, they will develop a job specification or "spec" and then begin to identify candidates. They do this by tapping into their network, often consisting of people they have already placed along with incumbents in similar positions from competitors or adjacent industries. If one of them calls you for recommendations, you may ask for a copy of the spec, though on rare occasions, it may be confidential, especially if the incumbent does not know he is being replaced.

If you are interested in the job yourself, do not be terribly shy about saying so, though try not to be too eager. Often executive recruiters are seeking people who are capable, talented, and happy in their current roles. If the candidate needs to be wooed a bit, so much the better, but don't be too coy or they will not keep calling. If you aren't interested, it is also perfectly OK to say so, but make sure you have good reasons. For example, you can say "I am leading a project right now that has several months to go before it is completed, and I have a strong personal and ethical commitment to see it through." If you say, "My daughter is a junior in high school and I won't move until she graduates," this is OK and certainly honest, though the recruiter may propose a commuting option for a proscribed period as an alternative.

You may say "I would never live in California, New Jersey, the Midwest, Hong Kong . . ." And while this may be true, you really ought to be more open and say something like, "You know, it would be hard for me to get terribly excited about opportunities in Detroit (Cleveland, Houston . . . whatever doesn't float your boat) but certainly for the right career opportunity, I would consider most anywhere." You can always turn the job down later if the work doesn't offset lifestyle sacrifices, but make this decision early in the process or else lots of people will get upset. It would be truly unethical to interview for a job you would never take, however. And like all business communications, you need to be completely honest, but especially so with the executive recruiters. They are extremely trustworthy as is necessitated by the nature of their work.

Assuming that, for one of several legitimate reasons, the job is not right for you; by all means, reach into your network and recommend people you truly respect and who are good matches for the job. A recruiter who finds that you are a good source of talent will not only appreciate your judgment and network, but will come back often with other opportunities. You want them calling often. Sooner or later, the job she describes will be your dream come true. Job placement at the executive level is a dance that takes time and patience. You must continuously build your capabilities and talent, your network and your resume before you will be considered as a serious candidate. By all means, always call back promptly if you miss their calls—they are your ambassadors.

Once you actually become a serious candidate, the executive search firm will carefully vet you including putting you through a number of tough interviews, along with fairly extensive background and reference scans. They are paid to deliver and they don't want the firm to come back in six months complaining after a painful and embarrassing separation and a costly renewed search. You should expect these interviews to be among your hardest. Screening candidates and separating the wheat from the chaff is their *raison d'etre*. Make sure you carefully review the questions in Chapter Three, and think hard about why you are really the right person for the job. Be prepared to give very crisp answers with concrete data. If you haven't made a significant measurable impact on your current and past organizations, you are not ready for the big time. Try to anticipate what they might ask you. Also, do lots of homework on the company with the opening. This is a big decision for everyone involved.

Generally the search firm will want to present a slate of candidates, but will often tip their hand to the company that one of their candidates is a true standout. The hiring company will always make their own, independent choice of whom they hire, but search firms are the experts in the business of identifying strong candidates for specific leadership positions. Even the staunchest executive who makes significant hires a couple times a year will have trouble second-guessing a search firm whom they have paid six figures for their services and opinions. In other words, if one candidate is highly recommended, she is likely to be the one the firm eventually hires.

A few tips. You can send your information to recruiters, but a referral is so much better. It is OK to speak with your executive friends, mentors, and sponsors to solicit their help in getting on the recruiters' screens. Some have websites that allow you to enter your bio and interests. These are OK and probably don't hurt, but realistically, executive placement is more often word-of-mouth from trusted sources.

When speaking with executive recruiters, as always, be completely honest and open. Others may disagree with this advice, but I always told them exactly what my current compensation was. You might even be asked for proof of current compensation, especially if your current comp is out of line for the industry. It is important to be realistic about what your switching costs will be—i.e. what it will take to get you to take a chance on something new. You ought to have a very good idea what similar positions are paying, and you should get good compensation advice from the search firm. After all, they want to consummate the deal and don't want you balking on a low-ball offer. Nor do they want to inflate your expectations only to have you walk away, disappointed. It is fine to negotiate for a bigger package, but as always, remember that "pigs eat; hogs get slaughtered."

I almost "fouled the nest" during negotiations for my first officer job. I was giving up a fortune in unvested options, pensions, deferred executive compensation, and even annual vacation time that I had accrued at GE. Working with the executive recruiter as an intermediary, we reached a serious impasse in the negotiations when I crossed the line with my demands to the hiring president, and the job offer almost fell through. I called him up directly and explained my concerns about giving up so much, but that I recognized he couldn't make me completely whole. I told him I still wanted the job, and that if there were no hard feelings, I would seek more common ground. The direct approach worked as I eventually took the job. I felt I was making a terrible financial decision but I had solace that the quality of the job would be fantastic and worth the sacrifice. Boy was I ever wrong about the financials. The job was truly outstanding, *and* working for the next ten years at the officer level paid me way more than I would have had staying with my former employer at the lower level. Yes I gave up job security, but realistically, if you have talent and confidence, you should plan on succeeding in your new role, in which case, the security issue is moot; the world always has a place for successful, talented, experienced leaders. Go for it and the money will come.

Chapter Twenty-Four

Leadership, Strategy, and Change

THROUGHOUT YOUR CAREER THAT, HOPEFULLY, INCLUDES a sequence of progressively responsible positions, your role will also become more strategic. We talked about how "managers do things right and leaders do the right things." How do you go about doing the right things, since this invariably means driving changes?

Certainly your plate will be plenty full simply by leading larger, more complex organizations with more and more impactful outcomes. You will have more people with all their personal and organizational issues. You will have more projects with bigger budgets and more critical deliverables. It is a huge accomplishment just to get these done in some reasonable facsimile to the desired outcomes. But if you really want to take bigger steps, you will also need to become more strategic to complement your tactical approach. What do I mean by that?

Let's consider what I mean by tactical, first. To me, tactics are the things you do to get the job done—scheduling, budgeting, hiring the right people, obtaining the materiel resources, setting priorities, holding peoples' feet to the fire, and delivering on expectations. Strategy, on the other hand, is determining which jobs to do, and looking around corners to determine where your organization should be going. It is planting seeds for a greater future. It is innovation, driving change, and leadership.

As you move along your career, each position will likely afford you greater strategic opportunities, if you choose to take them. And it will certainly require initiative on your part. More than likely, your boss, the other executives, and the board will know that this is their role, and will be more than happy to fill the vacuum if nobody else does. So you will need to get in the habit of formulating good strategic ideas early on in your career, and then learn how to sell them up the line. This is a pretty daunting task for a lower-level employee or middle manager, since the expectations for you are mainly tactical—simply to complete the assignments that are given to you. But you also may be closer to the action than they are, and if you make good observations and think carefully about formulating a winning strategy, you may be the ideal person to drive your company to a better place. More than likely, your leadership team will quickly recognize that you are special—a bigger thinker with ideas, an observation that could position you to move up faster.

So how do you formulate good ideas? A good place to begin is to understand your organization's current strategy, which is probably pretty good, assuming you are successful and well-led. Ask yourself, "What are the products or services we provide? Who are our customers? Who are our competitors? Who are our suppliers? What is working well and not so well? Are we on the upswing, flattening, or on the downswing? What is the trajectory of our industry? What is the global situation? How vulnerable are we? What new technologies or demographics might impact us?" It is more than OK to discuss these questions with your brass, especially if they have open door policies or skip-level meetings. Really you ought to be curious about these questions since they are so vital to your future. If you aren't, maybe you are in the wrong job, frankly. Actually, I have found that most people are more concerned with getting their specific work done and less so with the big picture. Don't let you become one of them.

Like most topics, there are a number of great books you should read in this area. I know you are already a reader, or you wouldn't be here with me now, so you are way ahead of the crowd to begin with. Adding 'strategic-thinker' to your personal brand is a huge positive step. Here are a few must-read books. The seminal work on competitive strategy is brilliantly titled *Competitive Strategy* by Michael Porter. In this book, Porter systematically unravels the dynamics of competitive forces including the various stages of an industry, the evolution of competition, customer and supplier leverage, switching costs and so on. The book is not an easy read, but it is well worth your time to read, and re-read it often. Another great book is *Blue Ocean Strategy* by Kim and Moubourne. Here the authors discuss how to get away from competition and the bloody red ocean of the marketplace battles, by sailing into unchartered waters where there are no competitors and where you can enjoy first-mover advantage. They offer great examples and a systematic process to uncover these opportunities. Find one for your company and you will be a hero of epic proportions.

Let me give you a brief example of a Blue Ocean Strategy that some colleagues of mine came up with at Nalco. Nalco sells chemicals for preventing scale, corrosion, and fouling of cooling towers. These towers use cold water to cool various process equipment across several industries including chemical, manufacturing, petroleum refining, agricultural, and even municipal. Over time, scale from the salts in the water, slime from microorganisms, and corrosion occurs in these systems. Nalco and a host of competitors sell the chemicals to prevent these issues from getting out of hand. But cooling water is always evaporating, fresh water is constantly being added, spent water is let down and flushed away, and the process is cycling. So whatever chemicals that were initially added to the cooling water were either being consumed or lost and nobody ever knew exactly how much there was left, or even how much was needed to begin with. One of our chemists discovered that if you added a special dye with the chemicals, then you could use a simple fluorimeter to measure the concentration of the dye, and it would correspond directly to the concentration of the active chemicals. Simply by knowing how much active water treatment chemical was in the cooling water at any given time was a huge advantage. Quickly Nalco became the leader in water treatment as a result, was able to sell commodity chemicals at premium prices because of the added information component. And once you know how much is actually there, it is a smaller step to determining what the right amount is for a given system.

This tracing concept, with the trade name Trasar®, is an example of creating a sustainable competitive advantage. If you read and grasp these two books, *Competitive Strategy* and *Blue Ocean Strategy*, in the perspective of your own business and industry, you will have a great foundation to create a sustainable competitive advantage in your space. But remember that most of your executives, and those at your competitors, have also read and studied these books. What can you add that is new, different, better? Don't sell yourself short. As an engineer, researcher, and innovator, you may be closer to the action than they are, often coming from Finance, Sales, or Marketing, for example. You can see innovations around the horizon and combine them for breakthrough advantage. Nalco combined information and product to step out; can you do the same? Are there new materials or designs that can make your products perform better or in new ways? Think about how composites changed the automotive and aerospace industry; or superalloys and high performing ceramics in power generation and jet engines.

Very typically, organizations grow by beating the competition or by creating brand new value propositions, and these two books provide blueprints on how to do this. A third way for organizations to grow is by moving into adjacencies. An excellent treatise on adjacencies is *Beyond The Core* by Chris Vook. Vook recommends that small steps are often better than new ones, along either adjacent technologies and products or adjacent customers and markets. Big steps involving new markets, products, and customers are great when they work, but represent major risks and have lower success probabilities as well. Keep your eye out for adjacencies, especially ones that might not be obvious. For example, your company may be highly product-focused, but your world-class supply chain might be leveraged to new markets. Consider how Amazon migrated from books into consumer products and durable goods, for example. If you can take your basic product line and modify it in a way that it can serve new industries or applications, this is a great place to start.

As you become more innovating in your thinking, you will certainly want to read Clayton Christiansen's classic *Innovators Dilemma* in which he describes how difficult it is for incumbents to replace themselves or do something strategically different, given that their current strategy is already working. And you may be interested in Andy Grove's *Only The Paranoid Survive*, which describes how Intel actually made the big leap by exiting the memory chip space where they were leaders, and moving into processer chips where they were neophytes.

I cannot emphasize enough how important it is to read great books, not limited to strategy, but also about leadership, human resources and organizational development, finance, politics, self-help, influence, communication, selling, biography. If you want to accumulate a lifetime of experiences, you could wait quite a while until you are old and wise, but can you really wait that long? By then, it will be too late. But if you read a great book by an important someone who has already accumulated a lifetime of experiences, in just a few hours, you can know what they learned the hard way. Isn't that fantastic? Isn't the written word one of the great differentiators between animals and humans, that which allows us to create great cultures and vast civilizations? Remember what Harry Truman said: "Leaders are Readers."

You may feel you don't have enough time to read. After all, it is hard to sit at your desk with a book open when the boss is expecting you to be busy "doing stuff." And usually, there is more than enough stuff to do to keep you running from the moment you arrived early to the moment you left late. And aren't you just a little bit drained by the time you get home? Once home, there are the chores, the kids, the spouse— everybody demanding a piece of you. Who has time for a book?

I squeeze in reading in several ways. Just a few pages right before falling asleep is enough to get me through a book in a few weeks, and if the book is tough, it puts me to sleep right away. I always take along books when travelling, especially to read while waiting at the airport, or on planes or trains. I can't read in a car—it makes me carsick—but while driving, I really love books on tape/CD. The average person spends four hundred hours in their car each year. Instead of top forty tunes or talk radio, pop in a good book. You may think that you won't be able to concentrate as well as during dedicated reading, and you will be right about that. But as the old adage goes, 'half a loaf is better than none,' and you will still get a great deal out of a good book on tape, CD. If there are sections you really want to study in depth, you can always take the book out of the library and reread the key sections if need be. Plus, the speed of the spoken word is a fraction of most people's reading speed, so you may just find that listening works just fine. In a year, I will go through over a hundred books on tape while driving. My public library has a rich supply that I get for free. Books on CD are a great way to learn to speak a foreign language, complete with proper enunciation and accent.

These days, everyone is carrying a smart phone, replete with memory—enough to hold dozens of books. You can often transfer them directly, or use a service like www.audible.com. Then, instead of just tuning out when you jog or walk the dog, you can have a book going, using all that free time to absorb new knowledge.

As you fill your mind with the thoughts of the great leaders, you will be amazed at how often the lesson you just heard on the way home from work yesterday immediately applies to a situation you are facing today. Why rely on yourself to face an important issue when you can also lean on your personal team of advisors such as Peter Drucker, John Kotter, David C Maxwell, Zig Ziglar, Earl Nightingale, Brian Tracy, Mark McCormack and on and on.

Now that you are building your deep knowledge of the world of strategy, you can start to apply it to your immediate situation. You can ask, "Why are we doing such-and-such this way?" or "Why are we even doing this at all?" "What else should we be doing?" "If we weren't already in this business today, would we still want to be?" Then pull together a plan on what, and how, to do things differently.

Ideally, if you have the authority, you can sneak these changes in without asking for or needing any approval. You and your team of co-conspirators can test the waters quietly and see how it works. Maybe you can bootleg a new product or service or tweak one of your business processes. This is usually risky business, though you have surely heard that "forgiveness is easier to get than permission." Also, remember that bootleg projects, once they leave the confines of your walls, can open your organization to serious issues including liability, intellectual property, customer and supplier issues, product safety and so on. It is best to leave your independent forays to internal matters. We want courage, but not cowboys, spending our millions.

When you have developed a concept to the point that it will require buy-in from above, you will now need to put on your influencing/selling hat. Another great book to add to your list is *Influence*, by Robert Cialdini. Sadly, many great ideas are passed over as most people lean on the conservative side, afraid of risk and fearful of change. They may even be threatened by your initiative. Your job is to get them on board, not necessarily by extolling the virtues of the innovation per se', but more by what it can do for the person and organization that needs to allow you to take it to the next step, even if it means handing it over to somebody else. If this happens, it is still a good thing and you can take solace in the fact that your initiative is truly having an impact.

As you progress in your career, each stage will require a greater strategic element, and you will be judged as much on your vision and creativity as you are on your execution skills. Again, only through knowledge derived from reading and experience will you have the capability to lead change.

Finally, to really drive change, you will need not only good ideas, but also persuasive skills and courage. It takes a lot of courage to stick your neck out and propose new ideas. You invariably will get shot down, time and time again, until you get to the point where you may feel it is not worth it. This is the time your mettle gets tested. You must now put your head down and keep driving, and eventually something will stick and get adopted, or you will realize that you have grown bigger than your role or organization and will need to move on to someplace more amenable to your capabilities. As Oliver Goldsmith said, "Success consists of getting up one just more time than you fall."

Chapter Twenty-Five

When, and How to Quit a Job

When To Quit a Job

WHEN IS IT TIME TO QUIT your job and look for something better? How do you know when it is time to move on? Or are the things that bother you more about you, than they are about the job itself? In other words, if you are dissatisfied in your current job for some reason, is it likely that you will encounter the same issues no matter where you go? Lincoln said that "people are about as happy as they choose to be." Maybe you will be trading one set of problems for another.

Let's start with a couple basic questions: "Do you consider your current organization to be a good or bad place to work?" and "Do you foresee the kind of future you want in your current company?" Here are some of specific things to consider:

1. Do you believe in the mission of your organization? Are you doing good, important things that society needs and values? Are your products and services important? Does what you do matter? Are you proud to tell people what you do and where you work? One thing to consider, in perspective, is that, for all the talk of the temporary downturn in the economy, a country's political stability rests on a firm foundation of economic growth and wealth creation. Any

company that is providing useful goods and services to the economy is also creating jobs, paying taxes, and employing suppliers and distributors. Remember that anyone consuming your goods and services is doing so because it provides him value. Try to never lose perspective on why your organization, however small or large, plays an important role in the economy and society. If your company is making money, you must be meeting somebody's needs, and that is intrinsically good.

2. Do you find your business to be interesting and fascinating? Does it pique your curiosity or challenge your capability? Don't you just love a delicious, challenging problem to address, and does your organization provide an endless supply? Are you growing and learning every day?

3. How are your coworkers? Are they decent, hardworking people? Do you have the usual mixed bag of characters that you will find everywhere? Are there exceptional people from whom you can learn? Do you have a couple really good, close friends at the office? We all have unique perspectives on what is going on and whom we work with. People are always imperfect and enigmatic—often irrational and confusing. Some of your coworkers don't necessarily share your values or culture. Put any two people together in close proximity for a length of time and there is bound to be conflict. Surely there are always associates who are self-centered and ambitious. Others may be less than ethical and you wonder why management doesn't see it or doesn't do anything about it. Maybe there are people you dislike or distrust to the point of having to extricate yourself from the situation. Before you bail entirely, try to get yourself away from colleagues you can no longer tolerate by first moving somewhere else in the company, if possible. But if this keeps happening to you no matter where you work, maybe *you* are the problem, instead of the other guy or situation. In general, I try seeing the vast majority of people as good, bright, honest, fair, helpful, well-meaning,

trustworthy folks who are trying to succeed for themselves and for the team. People generally live up to, or down to your expectations, so if you face the world with suspicion and distrust, you will likely get that back in return no matter where you work.

4. Is your company successful and will it continue to be? Don't you prefer to work for a winner? Winners have resources to invest for growth, and tend to build on positive momentum. Winners offer opportunities for personal growth, promotion, and job security. Winners have money to reward employees and investors. If you are worried about your company and industry taking a downturn, perhaps you are not working for a winner. But maybe you are just the nervous type who sees dark clouds in the horizon and has no confidence that your team will face them, adapt, and still come out a winner. Before you leave for greener pastures, make sure the issues are real and permanent as opposed to just the usual grousing from the rumor mill.

    Success is often fleeting—*sic transit gloria* is the Latin expression for this. You will always need to continue to work hard and smart and diligently, but why struggle so hard on behalf of a hopeless loser? You may feel your company and industry is in a downturn, but is it permanent or transitory? As we saw in the last chapter, you can play a strategic role in shifting your organization's direction. Try that first before giving up. Often people see a temporary downturn and extrapolate it to an overall demise. The world is dynamic and can change quickly. One new product or market can flip things around, and before you know it, you are on a new growth trajectory. I would not jump ship unless I was pretty certain that there was no way to avoid impending disaster. I have worked for several organizations that were in trouble, but they found ways to grow despite globalization and commoditization, loss of intellectual property protection, severe and unethical competition, and even facing into new, superior disruptive technologies that could obsolete our own. It is amazing how fast organizations can evolve when they need to. Sometimes the changes are wrenching as divisions get divested and people get let go, but these situations teach you resilience and represent turnaround growth opportunities. Everyone wants to hire someone who has successfully navigated tough transitions.

Similarly, it is easy to look outside your organization and see what appear to be greater opportunities in growing industries. If you can catch a rising star, this can be a great ride for you. But as in investing, by the time you realize that a new star has been born, others may have gotten there first for the biggest rewards, and competition is lurking to get their share as well. On the other hand, as an insider in your industry, you may see emerging opportunities long before the crowd gets there, and a move can be the best thing in the world for you. My recommendation as always is to do your homework and due diligence carefully before you make a move like this.

If you are in a very precarious situation, you might consider testing the waters while you are still employed; it is generally much easier to get a new job while you are currently working. But also, many companies who are downsizing offer nice parting gifts including several weeks pay, unemployment payouts, outplacement and retraining. Sometimes waiting to get let go isn't so bad an alternative.

5. How is your total compensation—pay benefits, facilities, location? Chances are, you have some serious switching costs including relocating and all the expense and hassle that goes along with a move. Our customers generally don't shift suppliers unless they see something like a twenty or thirty per cent benefit. Why not use that as a good rule of thumb for yourself? Are your work facilities safe, clean, modern? Do you have light and airy offices with ample space? Are you located in a nice part of nice town, close to affordable housing and good schools? How long and difficult is your daily

commute? Are you nearby recreational and entertainment activities that you enjoy? Do you or your family members require special educational or medical facilities or have limited employability elsewhere?

6. What is your career trajectory? Do you envision a future for yourself that you want? Are you stuck or moving along too slowly? Have you stopped growing? Have several others passed you by, especially those you consider less capable? Have you received a career-limiting performance appraisal or committed a *faux pas* that you will never recover from? Has your supervisor made comments that indicate you are on your way out? Do you feel progressively distanced from key decisions and decision makers?

    In Asia, there is an important concept known as 'face', which is a measure of one's prestige and power. A public disagreement and apparent humiliation can lead to loss of face in any culture, and could be an important signal for the employee to move on. For example, one of my organizations once held an important meeting for key leaders in the company. Lower level employees who weren't invited may have been inspired, asking themselves, "What do I have to do to get invited next year?" On the other hand, higher level employees who thought they should have been invited, but weren't, got a rude awakening and a major loss of face; surely they had to consider leaving the organization rather than facing the loss of face with their employees.

7. Are you stuck behind a supervisor who is going nowhere for a long, long time, but who won't help you make a move around him? Is your supervisor unethical or unfair, or just plain incompetent? John C. Maxwell brilliantly wrote in *The 21 Irrefutable Laws of Leadership* that a "5" or "6" will work for a "9" or "10," but not the other way around—at least not for long. One of the best moments I ever had during a performance review was when my supervisor said, "You

must be good. I've never seen so many people asking me if they could transfer into your group!" Look back at Chapter Seventeen to see how your boss stacks up.

8. Do you have an opportunity that is too good to pass up? Is the compensation a major step up? Are the responsibilities and personal growth huge? Does the move bring you a jump of a couple years from your current trajectory? Does it give you broadening opportunities in terms of new industry, region, market, technology? Does it bring you closer to your dreams? And does it scare you just a bit? Because if it is truly a step-out in growth, it ought to make you a little nervous about succeeding. If your future is a slam-dunk, it probably is not enough of a stretch.

———————

So back to the original question—when is it time to quit? I would say, when you are in a horrible, hopeless situation where you see no way out; where you have to drag yourself into the office each day; when you work with people, especially a supervisor, whom you can no longer trust and respect; when you and your company have limited foreseeable growth or future; when you have stopped caring about your work or coworkers or company; when you have been consistently unfairly treated, poorly appraised, badly compensated, have distressing work conditions; and obviously when you have a family obligation that mandates you relocate. Or conversely, when you have a seriously great opportunity that is simply too good to pass up.

In all of these situations, it is understandable that you would want to move on. I hope you will carefully and objectively scrutinize your current situation and the promises for something better to make sure you are not just trading one set of problems for another. If you are going to make a move, make sure it is for the right reasons and to the right place. Because I guarantee you that no matter where you go, there will be jerks, politics, unfair appraisals, frustration, tough competition and unreasonable customers—even if you are self-employed!

Timing a Move

In many economies and geographies and industries, you may experience a brief or prolonged tough time to be out there looking for another job. Maybe you are still employed and have put your search on hold, just waiting for things to turn around before you go out and test the waters.

In general, assuming some of the drivers, the ones we discussed above for making a move, are in play, you will be emotionally ready to consider a move. Often, however, great opportunities present themselves suddenly and without warning; this is a bit of a different situation that we will talk about in a moment.

I like to consider three factors in planning for a move.

First—is the work 'done'? Of course it never is done, but have you accomplished enough of what you wanted to? And is it left in a good place for you to leave? You certainly don't want to abandon a critical project at a critical moment. You have a responsibility to your coworkers and employer to make sure that this is a good time for you to move on. Of course, if you are an important contributor, there will always be a loss when you move on, but there are times in project when leaving would be unthinkable. For example, I once worked for a company that was making a huge investment in a new geography, including constructing major new facilities and making several key leadership and operational hires. Midway in the ramp up, the country manager quit for a better opportunity, and I felt this was inappropriate insofar as he was the key driver for making the investment. If you are at a critical moment and a dream opportunity arises, at least try to buy yourself some time to wrap things up. And frankly, there are some times when, what is good for your career should not come first. You do owe some loyalty to your employer, coworkers, and customers and quitting on them at a critical moment is unethical and unacceptable.

When I look at a resume and I see someone who has never left a job or company, I think to myself "tree-hugger" and that the individual is risk averse and not bold enough for my needs. On the other hand, if I see someone who has changed jobs more often that he has changed suits, I doubt the person will stay with my organization long enough to make an impact. Plus, I would be concerned that they have left their old responsibilities hanging and incomplete. On the other hand, a good resume shows some loyalty and longevity to an organization with a series of increasingly-responsible positions. And when they left one organization for another, there was a good compelling reason such as a major promotion or a major disruption in their industry or company such as an acquisition.

Next, in considering the timing of a move, have you groomed a successor and implemented systems that will allow the transition to occur without too much disruption? Sure you may feel that this shouldn't be your issue—that is your supervisor's problem. But remember, that is how the average person would see the situation and you want to be regarded as someone who is special—someone who is just as highly regarded after you leave.

Finally, do you really have something better to go to?

When those three things align—a good point in your projects to leave, people and systems to fill your shoes, and a great new opportunity— now it is the right time to go.

In one's career, there is certainly the preparation for a new job through formal education and on-the-job experience; there is certainly the "doing", successfully executing the responsibilities the job entails; and finally, there are the transitions between jobs. You need to get all three right or you will be limited in your opportunities. Make sure you treat transitions with as much respect as you do with preparation and execution. While I am loathe to use too many sports analogies, here you might think about the importance of playing offense and defense, but if you forget about special teams—the plays and players who participate in the transitions between offense and defense, you will often pay the price.

A Once in a Lifetime Opportunity

Every now and then and clear out of the blue, you might get that once-in-a-lifetime unsolicited opportunity to take your career and life in a major new direction. I recommend you take these chances very seriously, since they are so unusual. Big, bold moves give you the greatest opportunity for growth. You may not feel you have the requisite skills and experiences to succeed in the new job, but obviously, someone else does. It has been my experience that people surprise themselves with how much and how quickly they are able to grow into a major new role. Do not let a lack of self-confidence dissuade you from a big, bold move. I understand that there is a massive amount of upheaval associated with a largely unplanned and unconsidered move, especially one that is way beyond your projected career trajectory. My inclination is to go for it; you can always go back if it doesn't work out. And if it doesn't, try to make the return quickly; most people become stale after leaving a job or industry for more than a year or two. If you pass on this once-in-a-lifetime change, you may later experience regret and bitterness.

How To Quit a Job

This is another delicate situation that I have seen so many people mismanage. In general, I feel that a person's real character comes out after they have decided to quit. The best of people leave no "messes" behind, both physical and organizational, especially if they are leaving the job involuntarily, or are bitter and feel like they were mistreated and unappreciated by their employer.

By physical messes, I mean literally leaving their office and work areas clean and neat. Make sure you toss all your trash, personal memento's, clutter, and items that will be useless to your successor. Clean your area with the basic consideration that this would be a space you would like to be moving into if you were your own successor.

In terms of organizational messes, what I mean is to make sure that you tie up all loose ends before you part. Make sure that any pending responsibilities are carefully explained to your supervisor and successor so that they don't fall through the cracks. You want to train your successor and provide him, with encouragement to reach out to you, providing your personal contact information subsequent to your leave in order to make the transition smooth, regardless of the circumstances of your parting. As much as you may resent an involuntary separation, now is the time to really show class and character—everyone will notice and be far more inclined to help you later.

Even for internal transfers, this same mindset should apply. I have seen people transfer to a new role and simply drop everything, leaving resentful others holding the bag. If you do this, bad memories of you will circulate the company forever. As anxious as you may be to 'hit the ground running' on your new job and make a good impression with your new manager, even if it means working two jobs for a while, actively make sure the old job continues successfully after you part. You should not secretly desire for your organization to struggle without you; actually a truly great leader will be able to leave without the organization skipping a beat, because she will have put in people and systems that transcend her.

How to Handle Counter-Offers

If you are leaving your organization for a better opportunity, I strongly discourage you from using this to leverage a better deal from your current employer. If you have been successful and are highly regarded, your company may approach you with a counter offer. You tacit thoughts might be, "Where the heck were you the last two years?" but you should simply say something like, "I am really flattered. Working here has been great. I have made a commitment to my new employer, and I hope you will respect me as someone who keeps his commitments. If after a year or two, I may find things haven't worked out the way I thought they would, and I would truly be honored if you would leave the door open for me to come back."

People who do actually manipulate counteroffers make both parties angry and distrustful. After all, your current employer may feel like it is being extorted and your new one thought you had a handshake deal. Your recruiter will probably put your picture on her dartboard. And forget about much of a future at either place no matter which offer you take. You will be forever tinged with distrust. It is just bad business to go back on your word.

If you feel you have to give your current employer a chance to save you, then you must, early on in the discussions with your recruiter or future employer, tell them that, "I feel honor-bound to give my current employer the right to match any outside offers." This will sort of get you off the ethical hook, but it won't sit too well with anyone, regardless of the outcome, and will probably be a deal-breaker going forward. In general, my advice on counter-offers, solicited or otherwise, is to just steer clear. If you are ready to move on for the reasons we discussed above, then just go, and don't look back.

If you truly had a misperception about your future, shame on you and your employer for being less than candid and forthcoming. Every employee has a right to understand his career trajectory, and every employer has an obligation to ensure this happens. Of course, a particularly dirty secret of most employers is that they hate candor because most employees don't have the kind of future they really want at their current organization, and a candid discussion might just push them out the door, creating temporary problems for everyone.

It may be possible to avoid misperceptions altogether if you have a trusting relationship with your supervisor. In this case, you may feel comfortable telling her that you are testing the waters outside in order to accelerate your growth, career, compensation, improve your geography or whatever the real reasons are. Yes there is a risk that your employer will consider you disloyal and stop investing in you further, but probably your boss is also out there testing the waters *sub rosa* herself and will understand your very normal behavior. If she really wants to keep you, she will find a way to pave your future without a change of ownership. I would estimate that a large majority of employed people are almost always actively or passively exploring outside opportunities.

Summary

Most people make several transitions throughout their careers. Make sure you have really good reasons to make the changes, and recognize that change means growth and this is intrinsically good for you. You need to develop transition skills as much as your actual work skills. Use these moments to grow your abilities, confidence, and character. Leave your old job with class and make a huge effort to make sure that your move causes minimal disruption at your old post. Maintain your network at your old jobs while moving into your new ones. Understand that changes come with stress on everyone—for you, your colleagues, friends, and family. Go out of your way to be understanding and supportive. Geographical relocations will compound the stress. Go into each move with confidence and enthusiasm, and work especially hard to get a good head-start on your new position.

Chapter Twenty-Six

When and How to Fire Someone

As we saw in the last chapter, there are good and bad times to leave a job, and there are good and bad ways to leave a job. Sometimes, you get let go and sometimes you leave voluntarily. In this chapter we will see about involuntary separation from the vantage point of the supervisor. In the next chapter, it will be you who faces involuntary separation.

Assuming you are a supervisor at some point in your career, it is highly likely you will let people go. It is one of the more challenging and unpleasant tasks, though it needn't be; in most cases, both the individual and the organization are better for the separation. On the other hand, if it isn't difficult for you to let someone go, then perhaps you lack sufficient empathy to be an effective supervisor.

Although each separation is unique, depending on the circumstances leading up to the parting of ways, there are a few things that all separations should include. I can think of no exceptions to the rule that you should include your human resources and legal departments in the process, assuming your organization is large enough to have them. Wrongful terminations abound, and awards in the hundreds of thousands of dollars are not uncommon. Usually litigation can be averted with proper planning on the part of professionals.

Occasionally, as a supervisor, you may feel a strong impulse to let someone go. Perhaps your employee has disappointed you with performance; maybe there was an ethical lapse. You may feel that a pattern of underperformance or a poisonous attitude have caused you to reach your limits. You may be angry over the employee who has cost your team time and money. You must resist an emotional knee-jerk desire to fire the employee. You are a leader, a supervisor, and you must always maintain a cool head. Here is where your partners in legal and HR can really save your bacon by allowing you to safely blow off steam without compromising your leadership or your organization's strict adherence to policies that protect individual employees, supervisors, and the corporation from the costs, distractions, and angst of wrongful dismissal.

And in the end, if your employee has truly as earned an exit pass, she will soon enough depart with the full benefits of due process, and with the dignity afforded all employees when supervisors behave with proper planning and cool confidence.

A special case of separation involves a reduction in force or job elimination due to economic conditions. Here, presumably the individuals involved may have been performing acceptably, but your company can no longer carry them on the payroll. In this case, you must be extremely rigorous and prescriptive in the criteria used to identify individuals targeted for layoff; the risk of discrimination charges are magnified in this situation.

One of my organizations had a RIF (reduction in force). Each of the managers performed the "lifeboat drill" and which he found those employees that would create the least disruption as a result of their departure. About ten people were identified and we were well into our planned-process when someone noticed that most of the identified employees were Asian-Americans. Frankly, my recollection was that the slate was not-at-all the result of discrimination, but rather an unusual confluence of company needs and individual capabilities and experience.

Asian-Americans are so common in an R&D environment that we were well along the layoff process before anyone even recognized there was an issue. The leadership team quickly regrouped and made a more racially balanced slate and avoided the appearance of discrimination.

You may feel this is a case of reverse-discrimination, and perhaps it is, given that the people who were laid off were not necessarily the most deserving according to our own objective performance, need, and skill criteria; but pragmatically, the company avoided a costly and potentially embarrassing situation.

A common adage for successful leaders is "pick your battles carefully." Strengthening your organization by replacing weaker employees with stronger ones is certainly a battle worth fighting for, but make sure you get alignment from all of the organizations that may be affected and be willing to compromise and understand when things proceed more slowly than you might like. Your dud may be someone else's star, and your colleagues can make life very difficult for you when you remove the one person they feel they needed to be successful. Don't allow yourself to be set up as a scapegoat.

I will say that, under the atmosphere of cooperation and understanding, I have worked with many courageous HR and legal departments who have allowed me to remove, for example, a white male over 55 who had decided that he no longer needed to contribute, and a minority female over 40 who was simply over her head on the job and was earnestly trying very hard, but simply could not effectively perform the job she was assigned. As a footnote to the last case, her manager was a staunch defender and fought her dismissal. However, once she was removed, he was able to replace her quickly, and immediately the new employee began to make a big positive difference to his team.

Assuming that you have identified an individual for separation, and have obtained the appropriate approvals, you will now lead a separation discussion with the employee. I hope you are allowed to have HR representatives present, though your organization may not include this in its policies. This is a very trying discussion, and I strongly recommend you practice in advance with a trusted associate.

Your objectives in the discussion are to make clear that an irrevocable decision to make a separation has taken place. You will need to make it clear what the separation agreement will include such as severance pay, benefits, timing, and process. Do not allow this to become a negotiation; the decision has been made and it is final. You have an obligation to speak candidly about how and why the decision was reached. You may receive vehement objections and arguments as to why you have mistakenly selected the employee for termination, and you will need to calmly acknowledge that you understand their feelings, but that you are not open to going back on your decision. Expect an emotional response as the individual is suddenly experiencing a multitude of feelings including failure, disappointment, fear, embarrassment, shame, regret, anger, frustration, loss of identity . . . the full gamut. And yes, some employees may overreact and break down and even become violent. You probably have a good feeling about which of your employees may be prone to angry outbursts, and who might be volatile enough to strike out. If you have company security, you should have them nearby and on alert.

It is generally a good idea to choose the location of the discussion such that the employee can freely vent without disrupting his coworkers, and also one that allows quick exits in the case of an over-reaction.

No matter the reasons for separation, it is important for the supervisor to afford the employee dignity. In the case of reduction in force, this is a lot easier than with termination for cause. But even in the latter case, as a caring supervisor, you ought to include a good and fair summary of the employee's valuable contributions over the years, their unique skills and knowledge, and confidence that they will land quickly and find an organization that is a better match for their capabilities and interests. However, you will need to be clear that no representative of the company will be allowed to help them in their search for new employment, nor will any details of their separation be discussed.

At this point, some companies will walk the employee to the exit, take their badge, phone, and credit cards, and send them on their way. If there are personal items, some companies allow the employee to return with an escort after hours to gather his things. Others will simply pack up personal items in boxes and send them to the employee's home. It is probably a bad idea, except in the case of a fairly large RIF, that the employees are allowed to return to work for period of time. On the other hand, some companies and countries have rules that require lengthy advance notice of layoffs, and employees may continue on their jobs for several weeks before the actual separation. You probably ought to let the employee take the remainder of the day or week off, and return with a more positive attitude. Remind them that how they behave is critical to their reputation, their severance, and that there will be no tolerance for shenanigans.

After the employee is off site, of course the remaining employees will quickly learn that so-and-so just got let go. And they will speculate as to the cause. In a vacuum of information, the speculation can run rampant and take twists and distortions that disrupt morale and create fear.

As a supervisor, your hands may be tied as to what you can say about the departure of your former employee. But you must say something. At a minimum, you should pass the word that John or Mary is no longer an employee here, especially to people and projects that were counting on them. Ideally, you can discuss some of the circumstances leading up to the decision. For example, if there was a company layoff, you can be quite specific about the financial cause, the total numbers, the decision-making process, the new redistribution of assignments, and especially that there are no more layoffs planned.

If an employee was removed for cause, the situation is dicey. Frankly, I would relish the opportunity to share when an employee is removed for unethical or illegal behaviors. After all, if Joey was stealing or cheating on his time card and got let go as a result, this is a good lesson for everyone. But chances are that Legal and HR will insist on the code of silence lest the company risks a defamation suit. About the best you can hope for is that rumors come to you that will include some of the right reasons and some wrong ones. You can categorically, perhaps even laughably dismiss the wrong and ridiculous rumors and hope the real message gets out without your being explicit. I know this sounds like a cop-out, but as I say, this is a dicey situation necessary to protect both the company and the dignity of the person who was dismissed.

When I was let go as part of a change in ownership, I actually forget what I was told by the supervisor, other than about the benefit package, but I recall that the words around my leaving sounded rather hollow and insincere. I suspect that my boss' words were prescribed by HR and Legal and that his hands were tied. I was not terribly surprised by the dismissal as most of the officers at my level were going out one by one. And while I was escorted out by Security, I was allowed to return that evening to retrieve my personal effects. Sadly, I was not allowed to speak with my former employees to wish them luck and say good-bye, even though I was a company officer. I have no idea how my departure was described or how it was taken.

My dismissal, completely impersonal and unrelated to my performance, still left me unnecessarily with feelings of anger and disappointment. All the company really needed to do was say something like, "You understand that our new private-equity owners are slashing costs like crazy, and you are getting let go as a result. You're a good guy and you did well while you were here, and I am sure you will continue to be successful elsewhere. Stay in touch." And then, give me a few hours to meet with my staff to say good bye and trust that I will be professional enough to not trash the company. Instead—I got the "perp walk" and years of unpleasant memories.

I understand that if someone is being let go for cause—either for poor performance or an ethical lapse, then yes the company will want that person off site immediately. After all, the company is clearly calling the terminated employee a "bad apple," and that rarely sits well. But a lack-of-work, a downsizing or restructuring is another story altogether. Often a company will want the cooperation, and maybe even the re-employment of the displaced individual someday. Never forget that the world is really a small place. Your former employee may turn up as a customer, competitor, supplier, a regulator or public official, an expert witness or even a plaintiff.

It certainly behooves you to treat departing employees well and with class and dignity on the way out. After all, you will probably soon be following in their footsteps! Let's see how to face that career transition in the next chapter.

Chapter Twenty Seven

Getting Fired, Retired, Outsourced, Downsized or Just Plain Quitting

ALAS, ALL GOOD THINGS MUST COME to an end, including your job and even your career. For many of us, ending one career begins a new career by choice, while others among us may be forced out of our careers by changes in the workplace including downsizing, rightsizing, globalization, obsolescence, and even discrimination.

Most people who get terminated tend to gravitate to something similar for obvious reasons. But in many cases, the termination is a result of permanent, structural changes in an industry. Your termination may portend the end of your current career and the start of a new one. Even if your industry still has somewhat palatable though waning opportunities, now is a good time to consider alternatives.

For some people, career-ending means retirement and a cozy life on the golf course or beach, more time with family or to volunteer or do non-profit work. Can you afford it? What sacrifices might you need to make to stop living off of a paycheck for the rest of your life?

Some people plan to work until they drop, but they eventually have to quit for health reasons including their own failing health or that of a family member who needs their care. If your retirement plan is to work until you die, you'd better rethink that strategy. It isn't always, or even often, your choice in the matter, even if you are self-employed.

Unlike normal job transitions between relatively adjacent positions, career changing in this context means really starting anew, for whatever reason. Having only recently "retired" from my R&D career to focus on writing, coaching, consulting and boards, I am certainly no expert in these matters, so I will keep my comments brief and speculative. I have, however, observed many others who have made these kinds of changes, and as usual, they come with a mixed bag of success.

Most commonly, I have seen several individuals forced out of jobs that they would prefer to have kept, often for financial reasons. Let's face it—most people need to work for the income to maintain their lifestyle. Sadly, it is human nature to not save and invest enough to become independent of having a job. I strongly encourage everyone to make saving and investing a habit, and to start as early as possible to leverage the power of compounding. You don't want to be a hostage to a job you hate, and financial independence gives you the freedom to choose whether, and when, and to leave. People like to accumulate wealth for a number of reasons—to consume, for prestige, for pride—but to me, the freedom to change or end a career that has run its course is the number one reason.

Bob Brinker, financial planner and host of *"Money Talk"* on the radio has coined the term 'Critical Mass,' which is to accumulate enough savings and investments that can spin off sufficient income such that you no longer need a job to live the way you want. There are several financial calculators that help you estimate your number, and you might find interesting, a book aptly entitled *The Number,* by Lee Eisenberg about the subject.

No matter what your situation, it is helpful to consider what would happen to you if your job, company, and industry were to suddenly disappear. You don't need to look too far into the past to see several examples—the typewriter, the floppy disk, cathode ray tube displays, tape recorders, silver-based film; I bet you can think of dozen's more that have been made obsolete or at least, have been severely curtailed by advances in technology or changes in society. What would you do if your industry went kaput, even with time to plan?

Some people become entrepreneurs. Perhaps you have an idea for a new business or want to turn a hobby into full time work. Maybe you will decide that owning a franchise is your ideal. Maybe you have specialized knowledge that you can parlay into a lucrative consulting business. Perhaps you are pretty handy and good with money, and are a good fit to manage real rental property. Maybe your skills—quantitative thinking, good written and verbal communication, persuasion, nurturing and teaching— may easily transfer into a new and very different career. This can be a period of upheaval and excitement and growth, if you choose to regard it this way; that is, an opportunity to open up a new chapter in your life.

Often, I see a rather different psyche regarding forced transitions— anger, fear, frustration, bitterness. Perhaps the individual foresaw the changes coming but chose to ignore them. Maybe they were surprised by being put out to pasture before they were ready and with little warning. Maybe they are afraid that they won't get any new opportunities and will be permanently unemployed, or that they are too old to jump into something new. No matter what your age, a bold change can be daunting, even if it is by your own choosing.

If you foresee a career-ending situation in your future, or even if you don't, I encourage you to take stock of your abilities and interests and to come up with a suite of alternative next steps. Because when it hits, you need a plan or two.

It is certainly hard to avoid negativity when you have been downsized, but remember what Michael Corleone said in *The Godfather*: "It's not personal; it's strictly business," and move on. Of course you have strengths and weaknesses, but don't dwell on all the "woulda, coulda, shoulda's" of your recent employment. It happens to everyone. As we saw in the last chapter, I too experienced job loss when my company was sold to a private equity group and all the VP's got let go. It wasn't fun, but I was lucky to have a great network and marketable skills at the time.

And as an engineer, scientist, or innovator, chances are excellent that your skills are current and needed. You don't survive long in technology if you don't keep up with advancements.

Harvey Mackay wrote a terrific, uplifting book entitled *We Got Fired, And It Is The Best Thing That Ever Happened To Us*, comprised of a series of vignettes by some fantastic people who got the axe at one point in their lives, and how it propelled them to greatness. If you have a plan, it will give you the confidence and proper attitude to take the next steps in your life, whatever they may be. Sure, take a moment for mourning and introspection, but by all means, get moving, and fast.

No matter what your plans, now is the time to put your network to work (See Chapter Twenty-one on Networking), whether you are seeking employment, funds, employees, customers, or just general advice and encouragement. If you happen know someone who is currently unemployed and looking, a call to offer support and encouragement can be a lifeline to someone who will never forget that you helped them in a time of need. You get friends for life this way.

Summary

To assume that you will start and finish your career in the same basic field, industry, and organization is probably unrealistic these days. Try to prepare yourself emotionally, intellectually, and financially for periodic upheaval. Your career will most likely consist of a handful of mini—careers that may seemingly appear to be quite different from one another, but where you are the common denominator. Consider these transitions as opportunities to grow and recharge. Look for ways to bring your unique knowledge, experience, and network to bear in new situations, and leverage the things that made you successful, while avoiding the mistakes and pitfalls that caused you problems. You are indeed much wiser than you were years ago, and can be a tremendous find for a new employer or industry that is willing to take a chance on you, especially if you remain energized and enthused about your future.

Chapter Twenty Eight

Final Thoughts

THERE ARE MANY THINGS THAT PROBABLY matter to you in this world—the love of a spouse, the children you brought into this world and raised to become good citizens and contributors to society in their own right, the good you brought unto others, and of course, the substance and legacy of your career.

How do you live your work life and what is the impact of your career on your colleagues in particular, and the world in general? No matter what stage of your career, you should always feel great about yourself as long as you have been doing honest work that matters to society, and have been working hard and with integrity. That already makes you special in my book. Never, ever compromise your integrity. This is your foundation and no job, no wealth, no prestige is worth sacrificing your character and reputation for. Always do the right thing and you will be richly rewarded with self-respect.

Some special ways you can make a difference include mentoring new employees, making salient contributions to society through the products and services you provide, and through innovations you may have introduced. Opportunities to 'find a better way each day' abound and improvements, large and small, propel the flywheel of society toward growth and wealth creation for all. As an engineer, scientist, or innovator, your surely look for these and make them happen as the primary purpose of your job.

Your ability to make impact and to lead and influence others is really a strong function of who you are and how you grow. You must commit yourself to lifelong learning through listening, observing, reading, thinking, and doing. As you personally grow, so does your value. The greater the changes you experience in your work and life, the greater your growth. Gravitate toward bolder opportunities—new assignments, industries, people, places—that give you the greatest growth and have the confidence that you will grow into new assignments and that you will succeed. If you are getting too comfortable, consider this an alarm to make changes to get back onto that higher growth trajectory.

Always enthusiastically and energetically perform the work that you have been assigned. You will be evaluated mostly on your accomplishments and impact, but your attitude and values will often influence not only what and how you meet objectives, but also, the perception of others. Keep a very good log of all of your achievements and impact, no matter how small or seemingly inconsequential. Make sure that your value to your organization is always several times your cost to them and don't feel that you are being exploited, even when the multiplier is large.

Throughout your education and career, you need to be continuously building your network of friends and associates, and to tend to it like a farmer who tends to his fields. Make a serious effort to stay in touch and actively seek out ways that you can help your associates in their endeavors; and do so without any expectations or *quid pro quo*. Your unsolicited help will come back to you several-fold.

Finally, stop and smell the roses along the way. Find ways to enjoy each moment. Embrace the joy of a delicious problem; tackle with passion a new assignment; revel in a new competitor who will make you try harder and become stronger. Feel the pain of each loss and the joy of each win and make sure you take time to celebrate those little and big steps.

Take pride and satisfaction in your personal growth and never stop learning and growing. Become an expert in many areas and a generalist in others.

Make great and lasting friends throughout your career and enjoy the satisfaction of their growth and progress.

Remember that your career is a lot more important than just making a buck; it is all about making a difference. Good luck.

About the Author

CLIFF SPIRO WAS BORN IN THE Bronx in 1954 and grew up in Willoughby, Ohio. Cliff received a Bachelor's of Science degree in Chemistry with honors from Stanford University in 1976. In 1980, Cliff was granted a PhD in chemistry from Caltech where he shared the Herbert Newby McCoy prize for the top thesis in chemistry and chemical engineering.

In 1980, Cliff joined GE's Central Research Department in Schenectady, NY where he worked as a research scientist and manager for 15 years, which included projects in energy and fuels, synthetic diamonds, and advanced metallurgical and ceramic coatings. In 1985, Cliff transferred to GE Silicones where he led the silicone rubber R&D section. In 1988, Cliff was named General Manager of GE's Halogen Lamp Engineering Department.

In 2001, Cliff left GE to become VP of Research and Development for Nalco, a world leader in water, energy, and paper chemicals where he led a global staff of seven hundred. After a private-equity takeover in 2003, Cliff left Nalco to become VP and eventually CTO of Cabot Microelectronics, the world leader in CMP—Chemical Mechanical Planarization—a critical process in the production of leading-edge semiconductors. While at Cabot Microelectronics, he helped establish new labs in Japan, Taiwan, Singapore, and Korea. Cliff left Cabot Microelectronics in 2011 to pursue writing, speaking, offering innovation workshops, consulting, career-coaching, and boards.

Cliff holds twenty-one US Patents and has published and presented well over one hundred technical papers to international audiences and universities. He recently published "R&D is War—and I've Got the Scars to Prove it," consisting of a series or R&D war stories and lessons-learned.

Cliff has served on four corporate boards of directors including Maxdem, Inc. of Duarte, CA; The Mississippi Polymer Technologies Corporation of Bay St. Louis, MS; Strategic Diagnostics (SDIX—NASDAQ) Corporation of Newark, DE, and Araca Inc. of Tucson, AZ. Cliff has served on three academic advisory boards at U. Chicago, Northwestern, and U. Arizona. Cliff also served for six years on the Naperville United Way Board of Directors, and was elected Commissioner of Niskayuna Fire District #2 where he served for 2 years.

Cliff was a volunteer firefighter at Niskayuna Fire District #2 and is currently a volunteer firefighter at the Southside Savannah Fire Department, Skidaway Island, Savannah. He has two grown children and lives with his partner, Linda Artley, in Savannah, GA.

Finally, Cliff Spiro welcomes your comments. His website is www. cliffspiro.com and he can be reached by email at cliff@cliffspiro.com.

www.ingramcontent.com/pod-product-compliance
Lightning Source LLC
Chambersburg PA
CBHW031837170526
45157CB00001B/328